Independent Review of the Army Combat Fitness Test

Summary of Key Findings and Recommendations

CHAITRA M. HARDISON, PAUL W. MAYBERRY, HEATHER KRULL,
CLAUDE MESSAN SETODJI, CHRISTINA PANIS, RODGER MADISON,
MARK SIMPSON, MARY AVRIETTE, MARK E. TOTTEN, JACQUELINE WONG

Prepared for the United States Army
Approved for public release; distribution unlimited

RAND ARROYO CENTER

For more information on this publication, visit **www.rand.org/t/RRA1825-1**.

Published by the RAND Corporation, Santa Monica, Calif.

Library of Congress Cataloging-in-Publication Data is available for this publication.

ISBN: 978-1-9774-0883-9

Cover: Sgt. 1st Class Chris Oposnow/U.S. Army.

About This Report

This report documents research and analysis conducted as part of a project entitled *Assessing Full-Scale Implementation of the Army Combat Fitness Test*, sponsored by the Assistant Secretary of the Army (Manpower and Reserve Affairs). The purpose of the project was to provide the Secretary of the Army and Chief of Staff of the Army with recommendations to ensure the successful, full-scale implementation of the Army Combat Fitness Test (ACFT) on April 1, 2022; evaluate the extent to which such factors as age, gender, race, ethnicity, equipment availability, Army component, location, and climate conditions might influence a soldier's score and ability to pass the ACFT at the Gold Standard (i.e., the minimum standard); evaluate the ACFT's impacts on soldier readiness (including injury prevention and recovery), performance, and manning in all three Army components: Regular Army, Army National Guard, and U.S. Army Reserve; and assess the extent to which the ACFT may affect recruitment, retention, promotions, and talent management.

This research was conducted within RAND Arroyo Center's Personnel, Training, and Health Program. RAND Arroyo Center, part of the RAND Corporation, is a federally funded research and development center (FFRDC) sponsored by the United States Army.

RAND operates under a "Federal-Wide Assurance" (FWA00003425) and complies with the *Code of Federal Regulations for the Protection of Human Subjects Under United States Law* (45 CFR 46), also known as "the Common Rule," as well as with the implementation guidance set forth in DoD Instruction 3216.02. As applicable, this compliance includes reviews and approvals by RAND's Institutional Review Board (the Human Subjects Protection Committee) and by the U.S. Army. The views of sources utilized in this report are solely their own and do not represent the official policy or position of DoD or the U.S. government.

Acknowledgments

We are grateful to many individuals in the U.S. Army who offered their time, insights, and support to this project. We wish to thank the Office of the Assistant Secretary of the Army (Manpower and Reserve Affairs) for sponsoring this project, especially Jeffrey Angers (Deputy Assistant Secretary of the Army (Military Personnel) in the Office of the Assistant Secretary of the Army [Manpower and Reserve Affairs]) for his support throughout the project. We appreciate our action officers, COL Sharon Moore and COL Angel Gonzalez, who responded to numerous requests for information, coordinated our access to multiple Army organizations and units, and facilitated continuous engagements with a variety of key stakeholders. Members of our working group and our senior leader oversight committee were critical in helping to ensure we had access to needed data, participants, subject-matter experts, and other resources. They also provided invaluable feedback and insights over the course of the project. We owe a thank you to the many individuals who participated in interviews and focus group discussions and the many representatives from U.S. Army Forces Command, the Army National Guard, and the U.S. Army Reserve who helped coordinate discussions with those individuals. Those conversations helped us better understand and interpret the diagnostic data and informed our findings and recommendations. We thank the many individuals in the Army who assisted in providing data sets and documentation that contributed to the work.

Within the RAND Corporation, we would like to thank Barbara Bicksler, who worked tirelessly to improve our briefings and report; Judy Mele, who spent many hours in RAND's cold room processing numerous data sets and data feeds for the study; Jordan Reimer, who kept track of all our data files, coordinated focus group discussions and meetings, and helped keep us all organized; and Christine DeMartini, who assisted in the data processing and produced data displays during our interim progress reports. We are grateful to our RAND reviewers, Sean Robson and Jim Hosek, who provided timely and insightful feedback on the work at numerous points over the course of the study. Finally, we appreciate the careful review of this report by Patrick "PT" Henry (Henry Consulting Group and former Assistant Secretary of the Army, Manpower and Reserve Affairs). The feedback we received from all three reviewers greatly improved the quality of the report.

Summary

The Army Combat Fitness Test (ACFT) is a six-event fitness test designed by the U.S. Army Center for Initial Military Training as a replacement for the Army's long-standing Army Physical Fitness Test (APFT). The two tests are different in several ways. The APFT contained three events: push-ups, sit-ups, and a two-mile run. The ACFT has been expanded to six events: maximum deadlift, standing power throw, hand release push-up, sprint-drag-carry, leg tuck or plank (soldier chooses one), and a two-mile run. The tests are scored differently, and the ACFT has been designed to achieve a broad set of purposes: ensuring the physical fitness needed for combat, reducing preventable injuries, and transforming Army fitness culture.

Since 2019, the Army has fielded the ACFT in a diagnostic phase, during which not-for-record data were collected. This phase allowed the Army to continue studying the ACFT, provided the force with time to familiarize itself with the test and to train and prepare for it, and afforded time for the Army to plan out policies for how best to implement the test as the Army's official fitness test.

During the diagnostic period, as the Army learned more about how the ACFT functions in the field, it made adjustments and further refined its plans on how the test will be used in personnel decisions. In 2020, the Army asked RAND Corporation researchers to provide the Secretary of the Army and Chief of Staff of the Army with recommendations to ensure the successful, full-scale implementation of the ACFT on April 1, 2022. This request was driven by the 2021 National Defense Authorization Act, which directed the Army to commission an independent assessment of the ACFT.

Responding to this request entailed a broad review, including the assessment of such topics as prior research conducted by the Army on fitness tests, how the ACFT is being administered, how well the force is prepared, and the impact of the ACFT on personnel groups of interest. To conduct this review, we undertook a multidimensional approach that involved (1) an evaluation of ACFT data gathered by the Army on the workforce, (2) interviews and discussions with members of the workforce and subject-matter experts, and (3) a review and assessment of Army analyses, plans, policies, and other guidance that is relevant to the ACFT, including the broad body of evidence used by the Army when designing the ACFT and other relevant studies conducted outside the Army.

Top-Level Findings

Using our review of the ACFT, exploration of Army data, and discussions with the workforce and Army leaders, we arrived at the following top-level findings:

- **The evidence base to support the ACFT is incomplete.** The Army has demonstrated support for some, but not all, aspects of the ACFT. The Army has gathered a wealth of evidence on the ACFT. But the evidence gathered so far is mixed in its support of some of the fitness events included in the test, and there are gaps in the evidence base that are important for the Army to fill.
- **ACFT scores collected during the diagnostic period show some groups failing at noticeably higher rates.** The biggest impacts are observed for women, but we also see differences in pass rates across components, with the U.S. Army Reserve and the Army National Guard lagging behind the Regular Army, and across military occupational specialties (MOSs). Although these differences do not necessarily mean the test is flawed, they need to be investigated. It is also important for the Army to consider potential implications for personnel management that would result from high failure rates.
- **Research in multiple military settings has shown that training can improve pass rates.** The Army rolled out equipment force-wide, started training soldiers to improve performance on the ACFT, and provided some access to Holistic Health and Fitness personnel and master fitness trainers. Research has shown that training programs have generally had a positive impact on enhancing soldier fitness, but greater effort is needed to better understand and properly design tailored efforts to address distinct Army circumstances and personnel categories.
- **The Army would benefit from a formal management structure to oversee refinements to the ACFT over time.** In addition to addressing the recommendations of this report, the Army will face new issues or identify complex challenges that will need to be addressed following the ACFT's implementation. The involvement and attention of senior leadership in guiding, resourcing, and monitoring all aspects of the process are key to successful and sustained institutional change.

Recommendations

Building on these key findings, we provide recommendations (listed in Box S.1) that identify actions for the Army as ACFT implementation continues.

The most pressing issue for the Army to address is the evidence shortfall. As we discuss in this report, the leg tuck and plank especially are not well-supported for use in predicting combat task performance or for preventing injuries. These events in particular need additional evidence if the Army plans to use them to inform personnel decisions. All events could benefit from additional predictive validity and standard-setting studies on broader and larger samples of both men and women for all three of the ACFT's purposes.

In addition, greater consideration should be given to which soldiers should be held to combat standards versus general health standards. The test's standards are focused largely on ensuring minimum levels of combat task performance, and it is not clear that all MOSs or all individuals in any particular MOS need to be held to these standards or that the trade-offs

BOX S.1

Top-Level Recommendations

1. **Address shortfalls in the ACFT evidence base.**
   a. Collect additional data to further explore validity findings by gender and to establish justifiable minimum standards on fitness events.
   b. Establish proper justification for why all ACFT events and minimum standards apply equally to all soldiers.
   c. Continually examine and assess personnel decision outcomes that are associated with minimum scores.
   d. Define and continually assess organizational progress toward fitness transformation and solicit perspectives from the total force.
2. **Consider ways to mitigate impacts on the workforce.**
   a. Change how the ACFT is scored to mitigate impacts and align requirements with job-specific physical demands.
   b. Review and update fitness policies to ensure that a mechanism is available to address exceptional cases or circumstances.
   c. Use data from all test-takers to establish fitness tier cut points.
   d. Collect and analyze data on the impacts on the workforce, including recruiting, retention, promotion, and unit readiness.
3. **Take steps to further support training improvements over time.**
   a. Phase in implementation to allow time for individuals to improve performance on specific events.
   b. Implement training bands to help soldiers train for the ACFT and increase pass rates.
   c. Ensure soldiers have access to ACFT-relevant training, equipment, and coaching.
4. **Institutionalize a formal senior-level management structure to guide and oversee ACFT implementation and use.**

that would result from holding them to those standards are acceptable. Because this test will ultimately be used for personnel actions that can affect soldiers' careers and there are large differences in how women are performing on the ACFT relative to men, the Army must make sure it has strong evidence showing that the events and minimum standards it is using are valid and needed for all soldiers who are affected by them.

The second issue addressed by our recommendations concerns potential impacts on the workforce if scores are factored into personnel decisions or actions. With the pass rates we observed, many soldiers would be failing if the test were instituted today, including soldiers who were viewed as being in otherwise good standing in the Army. Given this potential outcome, the Army may want to put mitigation strategies in place to anticipate and manage areas in which the impacts will be the largest, including MOSs that have low overall pass rates and specific populations, including women and those over 45 years of age, many of whom are in leadership roles. Approaches to mitigation should be informed by data (including information on the potential trade-offs that might be incurred for those MOSs that may be unlikely to encounter combat).

Training, equipment, and supports also are important to position soldiers for success, and these resources make up the third issue addressed by our recommendations. Ensuring that all soldiers have access to training and equipment will be a continuous process that is refined and improved over time. The Army has instituted some training and provided access to equipment for many soldiers already, but the Army should also continue making improvements to training and equipment access to realize the benefits as quickly as possible.

Our final recommendation is intended as a long-term approach to address the ACFT's implementation and use over time. The Army must continuously monitor the ACFT after its full-scale implementation to ensure the test is working as intended and to address the issues raised in this report—the evidence base, failure rates, and training—and others that will arise.

Contents

APPENDIXES

Figures, Tables, and Boxes

Figures

Tables

Boxes

CHAPTER ONE

Introduction

The Army Combat Fitness Test (ACFT) is a six-event fitness test designed by the U.S. Army Center for Initial Military Training (CIMT) as a replacement for the Army's long-standing Army Physical Fitness Test (APFT). The Army has undertaken a multiyear and force-wide effort to study the ACFT, to allow the force time to familiarize itself with the test and to train and prepare for it, and to plan out policies for how best to implement the test when it is put in place as the Army's official "for record" fitness test, which is planned to occur on April 1, 2022.

The rollout of the ACFT development phase began in fiscal year 2019, when the Army conducted a one-year field trial with soldiers from approximately 60 Regular Army units, and has continued since that initial testing period, with all soldiers required to take the ACFT at least once.[1] Although ACFT results from this testing period have not been used for record, the Army has continued this diagnostic phase to enable additional data-driven learning, refinement, and adaptation. During this diagnostic test period, the Army has learned more about how the ACFT is functioning in the field and has adjusted the test.

As part of its preparations for full implementation, the Army in fiscal year 2020 asked RAND Corporation researchers to provide the Secretary of the Army and Chief of Staff of the Army with recommendations to ensure the successful, full-scale implementation of the ACFT on April 1, 2022. This request was driven by the 2021 National Defense Authorization Act, which halted the implementation of the ACFT pending an independent assessment of it, as follows:

> The Secretary of the Army may not implement the Army Combat Fitness Test until the Secretary receives results of a study, conducted for purposes of this section by an entity independent of the Department of Defense, on the following:
>
> (1) The extent, if any, to which the test would adversely impact members of the Army stationed or deployed to climates or areas with conditions that make prohibitive the conduct of outdoor physical training on a frequent or sustained basis.

[1] All soldiers were required to complete at least one ACFT by August 31, 2021. However, as of September 10, 2021, only a subset of the force had tested. In the Regular Army, approximately 70 percent of the enlisted workforce and 58 percent of the officer population had tested. In the Reserve and Guard, however, the proportion who had tested by that date was much lower. See Appendix A for more information.

(2) The extent, if any, to which the test would affect recruitment and retention in critical support military occupational specialties of the Army, such as medical personnel. (Pub. L. 116-283)

This report details the findings from our analysis.

How the New Fitness Test Is Different

The APFT has been the Army's fitness test of record for several decades. The administration and uses of the APFT have evolved over that time, but scoring and events have generally remained unchanged. The APFT consists of three events:

- push-ups
- sit-ups
- two-mile run.

For each event, a soldier's time on the two-mile run or number of sit-ups or push-ups completed are translated into a score of zero to 100, where 60 is a passing score for the event and a score of 100 reflects the highest score possible. Although men and women at all ages are issued a score from zero to 100 for each event and are required to score at least a 60, each event is scored differently depending on the soldier's gender and age (i.e., it is gender- and age-normed).

The Army identified the following purposes for the new fitness test: ensuring physical fitness needed for combat, reducing preventable injuries, and transforming Army fitness culture.[2] Accordingly, version 3.0 of the ACFT differs from the APFT in several notable ways.[3]

---

[2] The Army has articulated several purposes for the ACFT in its internal briefings, on its public-facing ACFT website, and in conversations with us. For example, as of December 2021, four purposes of the ACFT are listed on the Army's ACFT website: "improve soldier and unit readiness," "transform the Army's fitness culture," "reduce preventable injuries and attrition," and "enhance mental toughness and stamina" (U.S. Army, undated). However, in conversations with us, Army representatives generally identified the three overarching purposes we list in our report. For example, "enhance mental toughness and stamina" and "improve soldier and unit readiness" are encompassed in what we describe as the goal of *ensuring combat fitness*.

[3] The Army has made changes to the ACFT over the course of its pilot-testing with the workforce. Version 1.0 refers to the earliest version that was administered to personnel in 2019 and 2020. It designated three minimums (black, grey, and gold) depending on whether the soldier's military occupational specialty (MOS) had heavy, significant, or moderate physical demands.

Because of the coronavirus disease 2019 (COVID-19) pandemic–related social distancing requirements that limited the Army's ability to continue testing soldiers, and because soldiers struggled with passing the leg tuck, the Army launched a new version (ACFT 2.0) in mid-2020 (Brading, 2020). Changes that were instituted under ACFT 2.0 included the use of a two-minute plank for those who were unable to complete one leg tuck. The plank was not used in version 1.0.

First, the number, type, and sequence of events are different. The six ACFT 3.0 events are completed in the following order:

- maximum deadlift
- standing power throw
- hand release push-up
- sprint-drag-carry
- leg tuck or plank (soldier chooses one)
- two-mile run.

Second, the ACFT 3.0 is scored differently in comparison with the APFT. The ACFT 3.0 has the same raw score minimum requirements for all soldiers regardless of gender or age (i.e., the test is gender and age neutral, not gender- and age-normed like the APFT). Those gender- and age-neutral raw score minimums correspond to a score of 60 points on each event. All soldiers must score a minimum of 60 points on each event to pass the ACFT. This equates to a minimum of 360 points overall.

Research Questions and Approach

The Army asked us to explore the following research questions to inform Army senior leadership decisions about the implementation of the ACFT in fiscal year 2022:

- To what extent do factors such as age, gender, race, ethnicity, equipment availability, Army component, location, and weather conditions influence a soldier's score and ability to pass the ACFT?
- What are the ACFT's impacts on individual soldier readiness (including injury prevention and recovery), performance, and manning in all three Army components: Regular Army, Army National Guard (ARNG), and U.S. Army Reserve (USAR)?

Citing "lessons from ACFT 2.0," the Army announced the launch of ACFT 3.0 in April 2021 (Center for Initial Military Training, U.S. Army Training Command, 2021). ACFT 3.0 incorporated three changes. The first was the elimination of MOS-specific minimums (gold standard minimums were now applicable for all MOSs). Second, the plank was added as an optional alternative event to the leg tuck for everyone (under ACFT 2.0, it was only allowed for those who could not complete one leg tuck), and there was a change in how the plank was scored. The plank was now timed with a maximum of four minutes, and a new minimum time was set at two minutes and nine seconds (under ACFT 2.0, time was not recorded after two minutes, and two minutes was the minimum for passing). Third, gender-normed fitness tiers were introduced for future use in evaluations, order of merit lists, boards, and other personnel decisions.

ACFT 2.0 was in place when we started our work. ACFT 3.0 was rolled out in April 2021, while our study was underway. The ACFT 3.0 was still in place at the completion of our work.

- To what extent will the ACFT affect recruitment, retention, promotions, manning, and talent management, particularly in critical support occupations and in gender, age, race, and ethnic subgroups?

We used three approaches to answering these research questions: (1) analyzing ACFT data gathered by the Army during the ACFT's force-wide diagnostic testing period; (2) conducting interviews and discussions with Army leadership, members of the workforce, and subject-matter experts; and (3) reviewing Army research and supporting literature for the test's development, validation, and implementation.

We used all data available to us to answer as many of these questions as possible. However, because the ACFT has not been implemented for record and not all soldiers have tested, there were limits to which topics we could explore. Age, gender, component, and occupation data were generally available for all test-takers, so our analyses explore those variables in detail. We had some information on climate and elevation, and we conducted focus groups to gather anecdotal information about such topics as equipment availability, but these types of analyses were more limited. The effects of the ACFT on career outcomes and readiness cannot be known until the test is fully implemented and used in administrative actions. However, we discuss the ACFT's potential impacts and highlight areas of concern that the Army should monitor closely as the test moves forward.

Criteria for Judging a Test

Well-established professional guidelines describe what needs to be included in the evidence base that supports a test that will be used in an employment context, such as the ACFT.[4] In regard to best practices, important sources of evidence would show that the test

- predicts important outcomes (e.g., successful performance of warrior tasks and battle drills [WTBDs], injuries, or general health)
- is similarly predictive of the outcomes for all relevant groups (e.g., gender, race or ethnicity, age, or other groups of interest)
- is job-related for all who are held to its standards (e.g., performance on WTBDs should be mission critical for all occupations).

Performance differences across groups do not mean that a test is flawed per se, but such differences demand close scrutiny of the evidence to support the use of a test. If the evi-

[4] Those guidelines are summarized in two key documents (commonly referred to as the *Standards* and the *Principles*). See American Educational Research Association, American Psychological Association, and National Council on Measurement in Education, 2014; and Society for Industrial and Organizational Psychology, 2018.

dence shows that the events and minimums predict important outcomes and that they predict equally well for all subgroups of interest, then the test is considered *valid* for that use.[5]

Organization of This Report

This report contains top-level findings and recommendations from our independent assessment of the ACFT. In Chapters Two through Four, we explore three areas: (1) the evidence base for the ACFT; (2) variation in pass rates for different segments of the Army and over varying environmental conditions at the time of the test, the impacts of fitness tier cut points, and the potential impacts on the workforce; and (3) how training can affect ACFT pass rates. We offer findings and recommendations for each area. In Chapter Five, we summarize our key findings and recommendations and discuss an additional, overarching recommendation: the need to institutionalize ongoing oversight of the ACFT as it becomes the Army's official test of record.

---

5 Note that the best way to assess performance on a work task is to measure performance on the task itself. The ACFT events are not intended as direct measures of WTBD performance. Instead, they are intended as predictors of it. It is for that reason that we focus throughout this report on the importance of showing predictive relationships between the ACFT events and WTBDs. We also note that WTBD performance is not the only outcome of interest for the ACFT. Relationships for the other ACFT purposes (reducing preventable injuries and transforming Army fitness culture) should be demonstrated as well. See Appendix B for more discussion on this topic.

CHAPTER TWO

Evidence Base for ACFT Development and Validation

As the Army developed the ACFT, it amassed a large body of literature and conducted several studies to determine which events should be part of the test. This chapter summarizes our review of Army studies, which was supplemented with research from other relevant domains.

Stages for Reviewing ACFT Development and Validation

Our review was conducted using a six-stage framework of professionally recognized methods for developing, validating, and implementing physical assessments, as described in prior RAND work (Hardison, Hosek, and Bird, 2018). The stages, broadly defined, are (1) identify relevant performance criteria,[1] (2) identify potential predictor measures, (3) validate and select the best predictor measures, (4) establish minimum standards on predictor measures, (5) implement predictor measures in an operational setting, and (6) confirm predictor measures perform as intended. Because the ACFT has yet to be operationally implemented, we used findings from the Army's evidence base to make recommendations about how the Army can successfully implement and confirm predictor measures (stages 5 and 6) once the ACFT is operational. We examined all stages for the Army's three ACFT purposes: ensuring combat fitness, reducing preventable injuries, and transforming fitness culture.

Stage 1: Army Efforts to Identify Relevant Performance Criteria

This stage is critical because all follow-on analyses and decisions will ultimately point back to the appropriateness, value, or adequacy of the performance criteria. Appropriate criteria must be selected for each ACFT purpose.

To develop the criterion for *ensuring combat fitness*, the Army began with a comprehensive inventory of combat tasks (see Appendix C for more details on how this was done).[2] From

---

1 Criteria can be broader than just performance (e.g., injuries).

2 The Army's approach that we describe here was part of the Army's Baseline Soldier Physical Readiness Requirements Study (BSPRRS) as detailed in East, DeGroot, and Muraca-Grabowski, 2019.

this inventory, the Army research team conducted job analyses and selected tasks that were determined to be physically demanding, commonly occurring, and critical for all soldiers. The Army developed a WTBD simulation test (WTBD-ST) to assess the selected tasks, also listed in Appendix C. The WTBD-ST was administered to soldiers across various specialties to assess its measurement qualities. Subject-matter experts also reviewed the performance of soldiers on the WTBD-ST and determined the test to be a realistic measure of the individual physical fitness expected in a combat environment.

For the purpose of *reducing preventable injuries*, the Army developed a comprehensive taxonomy of injuries that are common to soldiers, as determined from medical data.[3] This classification characterized all soldier injuries regardless of cause or reason, whether on- or off-duty. Army medical surveillance reports from 2019 indicate that overuse musculoskeletal injuries represented 72 percent of all newly diagnosed injuries for active-duty soldiers (U.S. Army Public Health Center, 2020). Trends in the data indicate that women are injured at significantly higher rates than men. According to our review of the Army research and broader literature, this taxonomy is a viable foundation to define the injuries that are relevant to soldiers and to serve as a criterion for fitness event validation.

For the purpose of *transforming the fitness culture*, the Army assumes that a more diverse assessment will aid in changing individual and unit training behaviors. Accordingly, the Army designed the ACFT to be a comprehensive, full-body fitness assessment composed of a broad set of fitness components and fitness events.[4] This approach of training to a comprehensive fitness test has merit as a viable means of changing the physical fitness culture, but such an assumption must be explicitly measured and evaluated, which requires future analysis. We provide more details and a recommendation to this effect in the concluding section of this chapter.

Stage 2: Army Efforts to Identify Potential Predictor Measures

To support its selection of potential predictor measures, the Army conducted multiple literature reviews of military-relevant fitness components and events. These reviews included documenting how well each fitness component predicted performance on military tasks and each fitness component's association with reducing injury risk. Similar information also was examined for fitness events (if studies had been conducted at this level). The Army also collected primary data and analyzed secondary sources to identify fitness events within the components that had the greatest potential to be reliably measured and feasibly administered

[3] The Army's primary analytical work related to military injury prevention, which we reference throughout this report, is based on the following sources: de la Motte et al., 2017; de la Motte et al., 2019; and Lisman et al., 2017.

[4] Throughout this report, we use the term *fitness component* to refer to higher-level physical abilities (e.g., cardiovascular endurance, muscular strength, muscular endurance, power, and speed) and the term *fitness events* to refer to measures of those components (e.g., two-mile run, deadlift, hand release push-up, standing power throw, and sprints).

in large group settings. All steps of this work were informed by the perspectives of subject-matter experts (e.g., military operators and trainers, exercise physiologists, and measurement experts). The Army identified 23 potential predictor measures with multiple fitness events for each fitness component except for cardiovascular endurance. These 23 fitness events are listed in Appendix C.

Stage 3: Army Efforts to Validate and Select Best Predictor Measures

Ensuring Combat Fitness

The Army narrowed the 23 fitness events to a recommended set for predicting performance on the WTBD-ST using two progressive studies. This process resulted in a set of eight fitness events: the two-mile run, one-repetition maximum deadlift, push-up, leg tuck, power throw, sled drag, 300-yard shuttle, and sled push.[5] When the eight events were included in a single statistical model, the Army found that the events explained 74 percent of the variance in WTBD-ST completion times in the first study (conducted at Fort Riley) and 84 percent of the variance in the second study (conducted at Fort Benning).[6] Although the Army included all eight events in the regression equation, the results showed that the same level of prediction could be achieved using just four of those eight events (the sled drag, power throw, two-mile run, and one-repetition maximum deadlift).[7]

The level of prediction achieved for this new set of events was notably higher than that achieved for the APFT events. When the researchers created a regression equation using only

[5] The Army's technical report (East, DeGroot, and Muraca-Grabowski, 2019) does not detail the logic used in recommending all eight fitness events, but the initial determination for the inclusion of several events was based on statistical stepwise regression results. The original battery included seven events: sled drag, two-mile run, one-repetition maximum deadlift, sled push, push-up, kettlebell squat, and power throw. However, Army leadership expressed concern about a lack of full fitness component coverage. Accordingly, the Army research team substituted the shuttle run for the kettlebell squat and added an eighth event, the leg tuck.

[6] By *explained variance*, we mean the following: Individual soldiers completed the WTBD-ST and each of the eight events. The time it took to complete the WTBD-ST was recorded, along with the soldier's performance on each of the events. Across all soldiers, there was a range in performance: Some soldiers completed WTBD-STs more quickly than others, some soldiers achieved greater distance on the power throw, etc. In a statistical analysis, the Army examined how much of the variation in completion times on the WTBD-ST could be explained, or predicted, by soldiers' performance on the eight fitness events. In statistical terms, 0.74 at Fort Riley and 0.84 at Fort Benning were the R^2 values from the statistical regressions.

[7] See East, DeGroot, and Muraca-Grabowski, 2019, p. 52, Tables 19 and 20, which show that four events (the sled drag, power throw, two-mile run, and one-repetition maximum deadlift) explain 83.2 percent of the variance in WTBD-ST performance and that all eight events explain 83.5 percent of the variance (we note that R^2 always increases as more variables are added to the regression). Relatedly, the original seven events tested at Fort Riley explained the same amount of variance (73.7 percent) as the eight events (with the shuttle run replacing the kettlebell squat and the added leg tuck; 73.5 percent). See East, DeGroot, and Muraca-Grabowski, 2019, p. 48, Tables 14 and 15.

the three APFT events, they found that the APFT events explained only 42 percent of the variance in WTBD-ST completion times.

After the completion of the BSPRRS, Army leaders raised concerns about the recommended eight-event battery. The BSPRRS team conducted additional analyses to address the number of test events and administration time. The team's analyses resulted in the consolidation of three fitness events to save test time (combining portions of the sled drag, sled push, and shuttle run into a single sprint-drag-carry event) and in test modification to result in more-reliable event administration (e.g., converting push-ups to a hand release variant to determine successful performance more consistently). The drawback of these revisions is that the Army did not assess the validity of the modified or combined events. After these adjustments, the Army finalized the fitness battery to include six events: three-repetition maximum deadlift,[8] standing power throw, hand release push-ups, sprint-drag-carry, leg tuck, and two-mile run.

Our review of Army research and the broader literature for the leg tuck and plank (which was subsequently inserted as an alternative to the leg tuck) did not find supporting evidence for these events predicting physically demanding military tasks. The lack of details and documentation is concerning, given noted fitness performance differences between genders, especially since female soldiers perform significantly fewer leg tucks compared with male soldiers.[9]

The Army's technical report (East, DeGroot, and Muraca-Grabowski, 2019) presented statistical analyses of the extent to which soldier performance on ACFT events predicted WTBD-ST completion times, but the analyses did not assess whether there are differences by gender in the ability to predict WTBD-ST performance.[10] At our request, the Army conducted additional analyses of the data, including results for the ACFT fitness events (Appendix D contains the results). Our review of the Army's analysis suggests that there might be differences by gender in terms of how well certain ACFT events predict performance on the WTBD-ST (see Table 2.1).[11] Although the Fort Riley data set represents only a single study,

8 During the validation study at Forts Riley and Benning, the maximum deadlift event involved one repetition. The final battery involves three repetitions. Throughout the remainder of this report, we will refer to this event as *maximum deadlift*, but we mean the Army's current requirement in ACFT 3.0, which involves three repetitions.

9 The other ACFT events showed smaller differences between the performance of men and women than seen with the leg tuck. Neither the leg tuck nor the plank was well supported in either the Army's review of the broader literature or the Army's Fort Riley sample. It is for this reason we note that the leg tuck and plank need further validation support.

10 If there are differences in prediction by gender, it would suggest that there is *differential prediction*, which "refers to a finding where the best prediction equations and/or the standard errors of estimate are significantly different for different groups of examinees" (Young, 2001, p. 4).

11 If there are differences between men and women in the strength of the relationship between performance on ACFT events and performance on WTBD-ST, that would imply that the event (or test) may be more valid for one group than the other. This is one form of differential prediction, called *differential validity*, which

it is perhaps the most relevant study because of the direct applicability of the specific criterion, predictors, and sample relevance. However, its contribution to the larger body of ACFT validity evidence is limited by its small sample size, especially for women ($N = 46$). This small sample limits the confidence we can have in the results.

In addition, when we compared the original analysis to the gender-specific analyses, we determined that the original results predicted that women would do better than they actually do, on average.[12] This type of finding, in which different results are observed when men and women (or other subgroups) are combined in an analysis compared to when the analyses are run separately by gender, is important to examine and explore further.

Reducing Preventable Injuries

The Army conducted systematic reviews of research on the relationship between injury rates and broad physical fitness components. Overall, the Army's review of the research literature (much of which was conducted on military samples in basic training) showed strong to limited associations with musculoskeletal injury risk across the full range of physical fitness components, except agility.

With regard to the specific ACFT components and six associated events, the Army's review contained several injury-related findings. In studies with samples of sufficient size to report findings by gender, the timed distance run events were found to be strongly related to musculoskeletal injury risk for both men and women (i.e., soldiers who took longer to run a set distance tended to have higher injury rates, and vice versa). Push-ups were shown to be strongly supported for men but less so for women (i.e., fewer push-ups were related to higher injury rates). Sprint-drag-carry is a multidimensional event that measures speed, muscular endurance, and lower body muscular strength. Each of these fitness components had moderate relationships with injury risk (i.e., less fitness was related to higher injury rates).

Upper body power was found to have a limited association with injury rates and thereby provides limited support for the standing power throw as a predictor of injury (i.e., less upper body power is only slightly related to higher injury rates). The maximum deadlift as a measure of muscular strength was found to have a moderate relationship to musculoskeletal injury risk, but the risk of injury was higher both for those lifting the least weight and for those lifting the most weight (i.e., implying that either under- or over-training is bad). The leg tuck has not been specifically studied for the prevention of injuries. Core endurance has been found to be moderately related to injury risk (i.e., less core endurance is moderately related to higher injury rates). We note that findings that relate core endurance (typically measured

---

more formally "refers to a finding where the computed validity coefficients (e.g., predictor-criterion correlations) are significantly different for different groups of examinees" (Young, 2001, p. 4).

[12] In statistical terms, WTBD-ST was *overpredicted* for women. Formally, over- or underprediction "refers to a comparative finding where the use of a common prediction equation yields significantly different results for different groups of examinees. More specifically, overprediction means that the residuals . . . from a prediction equation based on a pooled sample are generally negative for a specific group, and underprediction means that the residuals are generally positive" (Young, 2001, p. 4).

with sit-ups) to injury reduction may not generalize to the leg tuck or plank.[13] We found no Army-specific analysis of planks relative to injury prevention.

Transforming Fitness Culture

Army analyses show that the ACFT—as a comprehensive fitness component assessment—supports the Army's efforts to transform its fitness culture via behavioral changes. Even during the diagnostic testing period, the yet-to-be fully implemented test changed soldiers' fitness behaviors, reallocated personal and unit training times, and improved specific and overall fitness levels because of sustained fitness training (see Grier et al., 2021). However, the Army has not systematically defined goals for transforming its fitness culture or established measures to assess progress for the diversity of environments and conditions reflected in the Army's total force.

Our Assessment of the Army's Validation Evidence Across Multiple Criteria

As noted in the prior section on criteria for judging a test, the Army contributes to the body of validation evidence when it demonstrates that a fitness event predicts important outcomes (criterion) and that such predictions are similar across various groups of interest. Findings from high-quality studies that involve sufficient sample sizes of relevant subjects contribute to greater confidence in the overall validity findings. Certainly, more validity evidence is better, but there are no absolute thresholds for the amount or magnitude of validity evidence. Consideration must be given to the relative magnitude and consistency of all validity findings across multiple criteria.

Table 2.1 provides a comparison of the magnitude and consistency of validation evidence used by the Army in its selection of fitness events for the ACFT according to the prediction of combat task performance and injury prevention.[14] The information summarized in the table comes from two types of sources: (1) the Army's primary research that shows relationships between ACFT events and the WTBD-ST (also referred to as the Fort Riley study) and (2) evidence from the broad research literature for physical fitness components. Detailed notes accompanying Table 2.1 describe the source of the reported information and the scales used by the respective study authors to categorize the levels of reported validation evidence.

For three fitness events—the maximum deadlift, standing power throw, and sprint-drag-carry—and their corresponding fitness components, the validity evidence is quite consistent across military tasks (in both the WTBD-ST study and the broader literature) and tends to be moderately related to injury risks (Table 2.1, Panel A). The estimates in the WTBD-ST study

[13] According to the Army's ACFT website (U.S. Army, undated), "the LTK [leg tuck] assesses the strength of the Soldiers [sic] grip, arm, shoulder, and trunk muscles." In other words, the leg tuck may not be a good measure of, and may not be intended to measure, core endurance. The Army's website also states that "the plank helps build core strength that promotes back health and helps reduce injuries."

[14] Transforming the fitness culture, although manifested through individuals, is associated more with organizational programs and leadership, as opposed to being related to individual characteristics, and therefore is not included in Table 2.1.

TABLE 2.1

Strength of Validity Evidence for Fitness Components and Fitness Events Against Multiple Criteria

Fitness Component Fitness Event	WTBD-ST (Fort Riley Study)[a] Men (N = 278)	Women (N = 46)	Systematic Reviews of Broad Literature Military Tasks[b]	Musculoskeletal Injury Risk[c]
Panel A: Consistent Evidence for Fitness Components and Events				
Muscular strength, lower body	–	–	**0.44**	Moderate
Maximum deadlift	−0.35	**−0.46**	–	–
Muscular strength, upper body	–	–	**0.46**	Limited
Standing power throw	−0.39	**−0.63**	–	–
Multiple fitness components[d]	–	–	**0.44 to 0.53**[d]	Strong to Moderate[d]
Sprint-drag-carry	**0.43**	**0.44**	–	–
Panel B: Inconsistent Evidence Between WTBD-ST Study and Broad Literature				
Muscular endurance, upper body	–	–	**0.49**	Moderate
Hand release push-up	−0.28	−0.19	–	Strong/Limited[e]
Cardiorespiratory endurance	–	–	**0.53**	Strong
Two-mile run	0.31	0.01	–	Strong
Multiple fitness components[f]	–	–	**0.45 to 0.46**[f]	Moderate[f]
Leg tuck	−0.23	−0.11	–	–
Panel C: Limited Study of Component, No Event Study with Army Soldiers				
Core endurance	–	–	0.34	Moderate
Plank	–	–	–	–

NOTE: A dash (–) indicates that no studies have been reviewed for the fitness component or event against the respective criterion: WTBD-ST, military tasks, or musculoskeletal injury risk. Hauschild et al., 2014, categorizes the strength of validity coefficients as follows: 0.70 to 1.0 is very strong, 0.50 to 0.69 is strong, 0.40 to 0.49 is moderate, and less than 0.40 is weak. Using this categorization, we bolded numbers in the table that are greater than 0.40 to highlight findings that are considered moderate or stronger.

[a] CIMT-provided analysis, June 2021. Numbers in these columns represent the validity coefficients of the WTBD-ST with each fitness event.

[b] Hauschild et al., 2014. Numbers in this column represent the average pooled correlations (validity coefficients) of all military tasks with each fitness component.

[c] de la Motte et al., 2017; de la Motte et al., 2019; and Lisman et al., 2017. The respective study investigators' ratings on strength of validity associations were based on the number of studies included, methodological quality, multivariate versus univariate findings, and magnitude of reported findings. See Lisman et al., 2017, for definitions of *strong*, *moderate*, and *limited*.

[d] Sprint-drag-carry reflects multiple fitness components: specifically, cardiorespiratory endurance, muscular endurance, and lower body muscular strength. The table shows the range of validity coefficients across the multiple components.

[e] Validation evidence for the hand release push-up and two-mile run with musculoskeletal injury risk reflects analyses of considerable research on the closely related APFT push-up and two-mile run. For the push-up, the results show strong evidence for men and limited evidence for women. For the two-mile run, the results show strong evidence for both men and women.

[f] The leg tuck reflects multiple fitness components: specifically, upper body muscular strength and core strength. The table shows the range of validity coefficients across the multiple components.

were slightly lower for the ACFT events than what has been observed in the research literature for the related physical fitness components. However, in two of the three events (maximum deadlift and standing power throw), women had higher correlations than men; the remaining estimates throughout the table were found not to be statistically different by gender.

In contrast, the next three fitness events (hand release push-up, two-mile run, and leg tuck; Table 2.1, Panel B) show inconsistent evidence between the WTBD-ST study and the broader literature. First, the WTBD-ST study showed a weak validity finding for the hand release push-up, whereas the systematic reviews determined the relationship between upper body muscular endurance assessments and military task performance to be moderate. For injury risk, the systematic reviews showed validity differences by gender for the push-up: strong for men and limited for women. Second, the systematic reviews for cardiorespiratory endurance found consistent and strong validities for both military tasks and injury risks. Conversely, the validity findings were weak for both men and women in the WTBD-ST study.

Finally, the findings for the leg tuck are quite complex and, according to the WTBD-ST study, indicate weak evidence that the event is valid, particularly for women. As a measure of core strength, there are confounding factors (e.g., limited upper body strength) that may affect the measurement of core strength for women. Therefore, women may have core strength that is not accurately measured because they lack the upper body strength required to perform a leg tuck. Our analysis of the Fort Riley data showed a strong correlation between the leg tucks and pull-ups (a general measure of upper body strength) and showed that this correlation was much stronger than the relationship between leg tucks and sit-ups (a general measure of core fitness).

Therefore, the original argument that the Army provided for including the leg tuck needs further justification, especially when some soldiers may lack sufficient upper body strength to perform a single repetition of this event. Also, in the Army's analysis, when the leg tuck was added, the ACFT's ability to predict performance on the WTBD-ST did not improve (as previously discussed). Considering these findings, the Army would need to perform additional work to demonstrate that the leg tuck measures core strength for all soldiers and that its addition to the battery provides value beyond existing measures, especially given the considerable performance differences by gender.

Finally, a void exists for validity evidence for the plank (Table 2.1, Panel C). According to conversations with the Army research team, our understanding is that performance on the plank has not been evaluated against the WTBD-ST.

In general, primary data collection and organization-specific validity research (meaning focused, targeted, and well-designed studies that capture an organization's specific environment and conditions expected for test use) are preferable when assessing a test and its respective application. In this case, the Army's research on the relationship between the ACFT (and its specific events) and performance on combat tasks is the best information to use when assessing validity. However, the Army's research should also be considered in conjunction with what has been documented about the test (or similar tests) in other contexts (by other services and in the broader literature). If the results of an organization's study are inconsistent with other published research, plausible explanations for such findings should be explicitly investigated. If the findings are not substantiated, replicated, or explained under properly conducted research, there is less confidence in the study's results.

That is the case for the Army's findings, as shown in Table 2.1, for the two-mile run and hand release push-up (particularly because of the previously mentioned concerns about the small sample size for the Army's validity study, especially for women). Conversely, although the systematic reviews show support for the core endurance fitness component as a predictor of military task performance, such research exclusively used sit-ups as the measure of core endurance: a fitness measure that differs from the leg tuck, which also requires upper body strength. In this case, there is no or only limited alignment between these respective events, and we must place greater confidence in the organization-level research to assess the appropriateness of the fitness event's use. As shown in Table 2.1, there was no evidence found for the leg tuck event in the Army's systematic reviews, and only weak validity evidence was found for the Fort Riley study in the prediction of WTBD-ST performance.

Stage 4: Army Efforts to Establish Minimum Standards on Predictor Measures

The Army employed a host of diverse means in determining minimum ACFT event standards. These methods included reviews of performance on criterion-referenced military tasks, actual soldier outcomes on WTBD-ST and fitness events, subject-matter expert judgments, and generalized statistical analysis and standards from other Army fitness tests. Additional methods and best practices for establishing minimum standards specific to the military fitness context are discussed in other RAND research (Hardison, Hosek, and Bird, 2018).

The Army established more evidence for minimum standards against the WTBD-ST than against criteria associated with the test's other purposes. More work is *potentially* needed for the purpose of injury prevention. The need for additional work depends on the Army's intended operational use of fitness scores relative to injuries. For example, the Army has historically used scores in the context of injury prevention or low fitness performance to identify soldiers' need for support services or programs. These uses result in relatively low-stakes decisions about which soldiers could benefit from such services. However, if the Army intends to use estimated injury risks to inform high-stakes personnel decisions (e.g., promotion, separation), more research would be needed to establish appropriate minimum fitness event scores.

Given the complexity of injury prevention and rehabilitation (McGill et al., 2015), it may be advisable for the Army to focus more on establishing a good foundation for standards by basing them predominantly on combat task performance. Injury prevention validation work then could focus more on supporting such decisions as which specific events help provide accurate assessments of soldiers' injury risks. In this manner, injury prevention does not need to be a primary driver in establishing ACFT minimum standards but rather can provide a distinct or supplementary perspective that is consistent with the Army's use of the fitness scores for this purpose.

One guiding principle for the Army is that its minimum standards should be periodically assessed relative to the performance outcomes achieved by various soldier demographics and the impacts on personnel decisions for similar groupings. The authority responsible for monitoring test outcomes and deciding on changes to ACFT standards was not evident from Army documentation, policy, or business practices. Such decisions involve equities spanning many Army organizations that often have competing interests. In Chapter Five, we recommend that the Army institutionalize a formal senior-level management structure to guide and oversee the ACFT's implementation and use.

Stage 5: Army Efforts to Implement Predictor Measures in an Operational Setting

The Army plans to decide on the ACFT for operational use by April 2022. Data collected during the interim diagnostic period and other ongoing Army analyses are being used to inform that decision. During the diagnostic period, two concerns were raised regarding ACFT findings: (1) the motivation of soldiers who took the test during the diagnostic period and (2) observed performance differences between groups of soldiers for some fitness events.

Testing under motivated conditions is essential to enhancing the precision of analyses to inform decisionmaking. Unfortunately, we were not able to obtain or collect primary data regarding this question because the test has not been administered in an operational setting. However, we reviewed lessons learned from implementations of past Army fitness measures that have implications for the full-scale ACFT implementation.[15]

The Army found persistent gender differences in fitness event scores in its analyses of test results during the diagnostic period. As part of ACFT 3.0, the Army in April 2021 introduced fitness tiers for future use in personnel decisions that require a physical fitness assessment score. Applying policy modifications may allow practical accommodations that address differences in fitness performance between men and women. Similarly, some research has shown that tailored and gender-based training may diminish the magnitude of gender performance differences in fitness events (Pierce et al., 2018). We discuss policy options to mitigate differences in pass rates in Chapter Three.

[15] The implementation of the Operational Physical Assessment Test and Indoor Obstacle Course Test, for example, suggests that the Army should prepare for changes in physical fitness requirements to be met with resistance, communicate frequently about changes in physical fitness requirements and the associated effects on personnel decisions, and provide proper instruction in training and allow soldiers proper time to prepare for the new physical fitness requirements.

Stage 6: Army Efforts to Confirm Predictor Measures Perform as Intended

Because the ACFT has yet to be fully implemented, we cannot determine whether it is working as intended. However, monitoring the ACFT—its measurement characteristics, processes, and impacts—requires a considerable commitment by the Army, as indicated previously.

Recommendations to Address Shortfalls in ACFT Evidence Base

The Army has conducted a variety of analyses and assembled a body of evidence generally consistent with professional guidelines that support the use of ACFT for its three purposes. Our review found that evidentiary support is thorough in certain areas and needs further work or data collection to confirm findings in others. The following are specific recommendations for how to address the areas that need further work.

Collect Additional Data to Further Explore Validity Findings by Gender and to Establish Justifiable Minimum Standards on Fitness Events

The Army's evidence base would benefit from additional research to properly demonstrate the ability of the ACFT events to predict both combat task performance and injury rates. The Army should seek to include more-representative samples of soldiers across the force (both physically and nonphysically demanding MOSs, as well as a range of ages), samples that include enough men and women across all fitness levels (including both the high and low ends of the fitness continuum) to be fully representative, and adequately large samples to explore validity separately by gender. This additional evidence is relevant to the Army's prior decision to incorporate the leg tuck and plank events into the battery. Additional evidence would provide further justification, or revisions if needed, for the Army's initial minimum standards that were established for each fitness event.

Establish Proper Justification for Why All Fitness Events and Minimum Standards Apply Equally to All Soldiers

Army policy unequivocally states that WTBDs apply to all soldiers. However, using information provided by the Army, we were not able to determine whether this policy should be applicable to a broad set of noncombat jobs. Other than a general "commander prerogative" relative to unit mission requirements or reference to predeployment training requirements, the Army was not able to detail persistent training and assessment programs in which all soldiers across all MOSs routinely trained to relevant physically demanding WTBD. To execute this recommendation, the Army should examine the fitness requirements needed to be successful in each Army specialty and consider the potential trade-offs among the associated costs, benefits, and impacts for periodically training to and achieving the requisite minimum fitness standards.

Continually Examine and Assess Personnel Decision Outcomes Associated with Current Minimum Scores

The Army set minimum standards under the guiding principle of establishing generally lenient cut points (tending toward not being overly restrictive in identifying acceptable performers).[16] This guidance also stated that minimum standards should be continually assessed to determine whether revisions are required. All standard-setting decisions and the outcomes associated with them involve a range of equities that include operational, force management, and policy considerations as well as the individual soldier's perspective.

The Army must be transparent about how such reviews will be conducted and how decisions will be made. This process should involve a diversity of operational, policy, and training perspectives, and data-driven analytical findings should be periodically reported to senior Army leadership for consideration in making needed adjustments.

Define and Continually Assess Organizational Progress Toward Fitness Transformation and Solicit Perspectives from the Total Force

The ACFT's purpose associated with transforming the fitness culture is only generally specified and lacks defined organizational expectations and measures to assess progress. The Army assumes that the broader dimensionality of the ACFT will drive behavioral change. However, as the Army seeks to transform its fitness culture, its organizational change approach should be a deliberate and explicitly focused effort that incorporates lessons learned from other successful implementations of large-scale Army initiatives. For this purpose, ACFT decisions, including data collection efforts and the design of supporting analyses, must be inclusive and representative of the diversity of the force and the implications of this diversity for performance outcomes.

To assess progress toward transforming the fitness culture, the Army should continue to gather information about soldiers' perceptions of fitness progress and commitments to training. Such analyses will allow the Army to evaluate the effectiveness of its many training programs and identify when new or more-tailored offerings are needed. The Army must also determine how to positively reinforce its cultural gains. As a final measure of cultural transformation, the Army must ensure that its fitness policies are current, relevant, and specific to the ACFT.

---

[16] The Army did not provide documentation for how it established the minimum standard on the plank event, so we are not able to review its methods for this event.

CHAPTER THREE

Variation in ACFT 3.0 Pass Rates, Fitness Tier Cut Points, and Potential Impacts on the Workforce

ACFT 3.0 policy requires that all soldiers, regardless of gender, MOS, or other characteristic, achieve a score of 60 on each event to pass. In addition, policy dictates the use of the ACFT total score to place soldiers into gender-normed fitness tiers.

In this chapter, we explore the variation in pass rates across different segments of the soldier population and across environmental conditions at the time of the test to help understand how current policy with respect to ACFT performance might affect different portions of the workforce. We also examine how different definitions of cut points for fitness tiers would affect the distribution of soldiers in those tiers. We then offer policy options that the Army could explore to mitigate the impact of differences in pass rates and in establishing fitness tiers.

Pass Rates Differ for Certain Segments of the Army for ACFT 3.0

The ACFT data that we analyzed for this study included ACFT scores gathered from soldiers across the force during the diagnostic period.[1] As presented in Table 3.1, ACFT pass rates differ between genders, across components, and between officer and enlisted personnel.[2] Most notably, scores for women are lower than scores for men across all components, and

---

[1] The results we present in this report include an individual's most recent ACFT score, which was entered into the system from 2019 through September 10, 2021. In 2019, the Army began administering the ACFT to a *subset of units* for diagnostic purposes. This continued through 2020. The Army then instructed *all soldiers* to take the ACFT at least once from October 1, 2020, through August 31, 2021. Many people who tested prior to October 1, 2020, retested per this guidance. However, many others did not. We include in our results only an individual's most recent score. In addition, we expanded our data to include data that were on record ten days after the deadline to allow extra time for data entry of scores at the end of the testing window.

[2] See Appendix A for information about the number and approximate portion of soldiers who have tested, by gender, component, and enlisted or officer status.

TABLE 3.1
ACFT Pass Rates, by Component

Personnel	Component	Men	Women	Overall
Enlisted	Regular Army	92%	52%	87%
	USAR	83%	41%	74%
	ARNG	83%	42%	76%
Officer	Regular Army	96%	72%	92%
	USAR	86%	49%	79%
	ARNG	91%	57%	87%

NOTE: Pass rates include the most recent test taken from 2019 through September 10, 2021. Those who completed an alternate event, were listed as being on profile, had missing data on one or more events, or had scores outside the expected range of scores on an event were excluded.

officers are passing at higher rates than enlisted personnel within components. These differences are most pronounced for women in the Regular Army (in which the officer pass rate is 72 percent, and the enlisted pass rate is 52 percent). In addition, pass rates in the Regular Army are higher than in the USAR and ARNG for both men and women.

Differences in pass rates were also observed across enlisted MOSs and officer areas of concentration (AOCs). As shown in Table 3.2, the top ten MOS pass rates for Regular Army enlisted women range from 89 percent to 65 percent; the bottom ten range from 44 percent to 31 percent. Even among the best performing MOSs, about one-third of the women who tested are not passing. However, it is important to keep in mind that not everyone has tested. In MOSs for which the share of women who have tested is small, the pass rates reported here could change significantly once everyone tests.

In comparison, pass rates for Regular Army enlisted men in the top ten MOSs are quite high, ranging from 100 percent to 98 percent (Table 3.3). Pass rates in the bottom ten range from 86 percent to 83 percent. The percentage of men who have tested is generally much higher than the percentage of women, so pass rates are less likely to change significantly once the entire population has tested.[3]

Table 3.4 shows the ten MOSs with the lowest pass rates overall, with gender-specific pass rates in those MOSs also displayed. These MOSs would be most affected by the ACFT pass/fail policies. If failures on the ACFT were used in personnel actions (e.g., eligibility for reenlistment, educational opportunities, select assignments, promotion decisions) and large proportions of individuals fail, retention rates in these MOSs may be hardest hit. That is, in addition to potential involuntary separations for failing the ACFT, soldiers in these MOSs

[3] However, the percentage of men (and women) who have tested in the reserve components is much lower than it is in the regular Army (30 percent of enlisted men in the USAR have tested, and 44 percent of enlisted men in the ARNG have tested, compared with 70 percent of enlisted men in the Regular Army). Therefore, we might expect pass rates in the reserve components to change significantly once more soldiers test. See Appendix A for more details.

TABLE 3.2

Top Ten and Bottom Ten Military Occupational Specialty Pass Rates Among Regular Army Enlisted Women

MOS	Proportion Passing	Number of Women in MOS with ACFT Scores[a]
Top Ten		
38B - CIVIL AFFAIRS SPECIALIST	89%	45
15U - CH-47 HELICOPTER REPAIRER	74%	62
09W - WO CANDIDATE	74%	80
09S - COMM OFF CANDIDATE	70%	191
94E - RADIO EQUIPMENT REPAIRER	69%	36
42R - ARMY BANDSPERSON	68%	94
12N - HORIZONTAL CONSTRUCTION ENGINEER	67%	149
19D - CAVALRY SCOUT	66%	89
31D - CID SPECIAL AGENT	66%	53
35M - HUMAN INTELLIGENCE COLLECTOR	65%	323
Bottom Ten		
68R - VETERINARY FOOD INSPECTION SPECIALIST	44%	234
94F - HOSP FOOD SERVICE SP	43%	37
13J - FIRE CONTROL SPECIALIST	43%	322
92G - CULINARY SPECIALIST	42%	1,381
68C - PRACTICAL NURSING SPECIALIST	42%	545
25V - CBT DOC/PROD SP	41%	87
91M - BRADLEY FIGHTING VEHICLE SYSTEM MAINTAINER	41%	51
68G - PATIENT ADMIN SPECIALIST	41%	167
25L - AN TSQ-73 OP/MAINT	39%	117
89A - AMMUNITION STOCK CONTROL/ ACCOUNTING SPECIALIST	31%	64

NOTE: Pass rates are only included in the table if at least 30 women in the MOS completed the test. Pass rates reported here include the most recent test taken from 2019 through September 10, 2021.

[a] Those who completed an alternate event, were listed as being on profile, had missing data on one or more events, or had scores outside the expected range of scores on an event were excluded.

TABLE 3.3

Top Ten and Bottom Ten Military Occupational Specialty Pass Rates Among Regular Army Enlisted Men

MOS	Proportion Passing	Number of Men in MOS with ACFT Scores[a]
Top Ten		
18Z - SPECIAL FORCES SENIOR SERGEANT	100%	221
18B - SF WEAPONS SERGEANT	100%	254
18C - SF ENGINEER SERGEANT	100%	239
18E - SF COMMO SERGEANT	100%	232
18D - SF MEDICAL SERGEANT	100%	227
12X - GENERAL ENGINEERING SUPERVISOR	99%	100
18F - SF ASST OP&INTEL SGT	99%	98
11Z - INFANTRY SENIOR SERGEANT	98%	1,183
38B - CIVIL AFFAIRS SPECIALIST	98%	716
09W - WO CANDIDATE	98%	840
Bottom Ten		
92S - SHOWER AND LAUNDRY SPECIALIST	86%	182
68Q - PHARMACY SPECIALIST	86%	290
51C - ACQ, LOG, & TECH CONTRACTING NCO	86%	148
92G - CULINARY SPECIALIST	85%	3,180
68X - BEHAVIORAL HEALTH SPECIALIST	85%	561
94S - PATRIOT SYSTEM REPAIRER	85%	212
68B - ORTHOPEDIC SPECIALIST	85%	106
94W - ELECTRONICS MAINT CHIEF	84%	96
94M - RADAR REPAIRER	84%	214
68M - NUTRITION CARE SPECIALIST	83%	196

NOTE: Pass rates are only included in the table if at least 30 men in the MOS completed the test. Pass rates reported here include the most recent test taken from 2019 through September 10, 2021.

[a] Those who completed an alternate event, were listed as being on profile, had missing data on one or more events, or had scores outside the expected range of scores on an event were excluded.

TABLE 3.4

ACFT Pass Rates for Bottom Ten Military Occupational Specialties Overall (Regular Army)

	Proportion Passing			Number in MOS with ACFT Scores[a]	
MOS	Overall	Men	Women	Men	Women
68T - ANIMAL CARE SPECIALIST	67%	92%	48%	177	242
68G - PATIENT ADMIN SPECIALIST	71%	92%	41%	234	167
68C - PRACTICAL NURSING SPECIALIST	71%	87%	42%	991	545
68M - NUTRITION CARE SPECIALIST	72%	83%	50%	196	100
68X - BEHAVIORAL HEALTH SPECIALIST	72%	85%	47%	561	288
92G - CULINARY SPECIALIST	72%	85%	42%	3,180	1,381
68Q - PHARMACY SPECIALIST	72%	86%	51%	290	189
68D - OPERATING ROOM SPECIALIST	73%	88%	50%	403	275
89A - AMMUNITION STOCK CONTROL/ ACCOUNTING SPECIALIST	74%	88%	31%	189	64
42A - HUMAN RESOURCES SPECIALIST	74%	89%	48%	4,146	2,402

NOTE: Pass rates are only included in the table if at least 30 men or 30 women in the MOS tested. Pass rates reported here include the most recent test taken from 2019 through September 10, 2021.

[a] Those who completed an alternate event, were listed as being on profile, had missing data on one or more events, or had scores outside the expected range of scores on an event were excluded.

might choose to voluntarily separate because of missed career opportunities. Many of the MOSs with the lowest pass rates are medical personnel, such as behavioral health specialists, pharmacy specialists, and operating room specialists, for which retention issues are well documented.[4]

Among officers, medical personnel also dominate the list of AOCs with the lowest pass rates; among male officers, command and unit chaplains have the second-lowest pass rate. See Appendix E for officer pass rates by AOC and gender.

We also observed meaningful differences in pass rates across age groups. As shown in Table 3.5, enlisted soldiers ages 18 to 44 have consistent pass rates regardless of component or

[4] Research has shown that all services, including the Army, have struggled to fill medical authorizations and retain medical personnel (U.S. Government Accountability Office [GAO], 2018; GAO, 2021). For example, GAO, 2021, examines staffing gaps (the difference between end strength and authorization) by skill level in enlisted medical occupations. The report states that 13 occupations had a staffing level below 90 percent in at least one higher skill level during the years studied, and four occupations had a staffing level below 80 percent. Behavioral health specialists, the MOS with the sixth lowest pass rate in the Regular Army enlisted population, had staffing levels below 90 percent in one skill level. The staffing shortages in these critical occupations may be exacerbated by ACFT requirements if policies are not put in place to mitigate the impacts of current pass rates. A more thorough analysis of pass rates, fill levels and retention rates, and job requirements by MOS can and should be conducted to support such decisions.

TABLE 3.5
ACFT Pass Rates for Enlisted Soldiers, by Age Group

Personnel	Component	18–24	25–34	35–44	45-Plus
Men	Regular Army	91%	93%	91%	86%
	USAR	86%	84%	81%	68%
	ARNG	85%	84%	81%	67%
Women	Regular Army	49%	57%	53%	40%
	USAR	42%	43%	39%	24%
	ARNG	42%	44%	39%	19%
Overall	Regular Army	85%	88%	87%	82%
	USAR	74%	75%	74%	63%
	ARNG	75%	78%	76%	63%

NOTE: Pass rates reported here include the most recent test taken from 2019 through September 10, 2021. Those who completed an alternate event, were listed as being on profile, had missing data on one or more events, or had scores outside the expected range of scores on an event were excluded.

gender. However, in every case, the pass rate for the 45-plus age group, which includes many of the Army's leaders, drops off relative to younger soldiers. The same patterns hold for officers (see Appendix E). Our examination of the relationship between age and individual event scores showed similar results, also provided in Appendix E.

As we discussed above, if we assume that people might leave the workforce at higher rates if their careers are adversely affected by the ACFT, the lower pass rates illustrated up to this point (for women, certain MOSs, and some age groups) have implications for retention among these populations.

Weather and Elevation Affect Performance on Some Events

The 2021 National Defense Authorization Act called for an analysis of the relationship between soldier performance on the ACFT and the weather and elevation at the time of the test (Pub. L. 116-283). We analyzed data on the following items to explain the relationship between test conditions and test scores, including total ACFT scores and individual event scores:

- individual characteristics (race and ethnicity, gender, age, height, weight)
- MOS at the three-digit level
- weather (the weather on the day of the test, including average temperature; presence of snow or rain; relative humidity; and average temperature across the month)
- elevation and region of the country.

Our results showed that higher elevation has a negative impact on performance on the two-mile run for both men (approximately five-point difference in score at higher elevations) and women (approximately ten-point difference in score at higher elevations). Temperatures over 90 degrees were also associated with slower times on the two-mile run (approximately five points for women and 2.5 points for men).

The results from our analysis need to be interpreted with caution. Several relevant factors are not included in our data, which could be confounding our results. For example, we do not know whether certain events were held indoors on days that were particularly hot outside or had inclement weather. If, for example, the plank was conducted indoors but the leg tuck was conducted outdoors, people may have opted to take the plank when they otherwise would not have.

Our data indicated that scores are higher on days when there was rain or snow, suggesting that those events might have taken place indoors on those days. In addition, results from our focus groups suggested that current policy requires additional weight to be added for the sprint-drag-carry when conducted on an indoor surface. Our data do not contain information on the type of surface on which an event was performed. These unknowns in the available data make interpreting the results of our analysis especially difficult. Tracking where events are held (indoors or outside, the type of surface) and any corresponding modifications to how the event was administered would provide information to control for these differences in future analyses.

Different Fitness Tier Cut Points Result in Different Outcomes for Women

The second element of the Army's ACFT 3.0 policy is the establishment of gender-normed fitness tiers.[5] Fitness tiers are constructed according to the distribution of total ACFT scores by gender and are intended to recognize top scores. The top tier, platinum, reflects the top 1 percent of scores. The next tier, gold, reflects those in the top 10 percent, and so on (silver = top 25 percent; bronze = top 50 percent; green = bottom 50 percent with a minimum score of 360). The Army has not yet determined exactly how fitness tiers will be used, but Army ACFT 3.0 policy indicates that fitness tiers may be used by promotion boards, for order of merit lists, and for other personnel decisions, which makes how these tiers are established particularly important.

We evaluated two approaches to establishing the fitness tier cut points: one in which the cut points would be normed based *only on those soldiers who pass the test* and the second based *on all soldiers who took the test*. Cut scores are calculated separately by gender. To

---

5 Although the Army refers to these as *evaluation tiers*, Army leadership indicated that these tiers may be used for purposes other than evaluation. We therefore refer to them as *fitness tiers* instead to acknowledge that broader use.

illustrate the implications of these two approaches on the workforce, Figure 3.1 shows the proportion of Regular Army enlisted women who would be assigned to each of the tiers for each approach.

Before we compare these approaches, we highlight two groups of women who are not eligible to be assigned a fitness tier because they did not pass the test. The first is the group who scored over 360 points on the ACFT but did not pass the test because they failed at least one event (if they scored less than 60 points on one event, but more than 60 on another, they could still achieve a score of 360). These individuals are shown with the relevant tier color according to their total score, but they are marked with a black outline around the bar, indicating they did not pass.[6] The second, larger group is those who are not eligible for a tier because they scored below 360 (shown in dark gray bars at the top of the figure). If failure is used to exclude personnel from fitness tiers and therefore career opportunities (e.g., promotions, reenlistments, or educational opportunities), the impacts on women would be large regardless of which tier approach is used.

Among those who passed the test (and thus are eligible for a tier), the cut points determine which tier they belong to. When the tiers are based solely on passers (Figure 3.1, top), a much smaller proportion of women would be in the platinum, gold, silver, and bronze tiers (which are considered desirable) than if all test-takers were included in the creation of the tiers (Figure 3.1, bottom). This difference is more pronounced in some MOSs than others (each bar in Figure 3.1 represents an MOS). To facilitate comparisons across figures, we have placed horizontal blue lines at the 40 percent and 25 percent marks. When looking at the "passers only" chart for women (Figure 3.1, top), most women are in the green tier at the 40 percent mark, with some still in the green tier even at the 25 percent mark. When the cut points are based on all testers (Figure 3.1, bottom), most women are at the bronze level at the 40 percent mark for most MOSs. The two approaches to norming have a negligible effect on the proportion of men in each tier, as shown in Appendix F. This is because men have much higher pass rates. Excluding the relatively small number of men who do not pass from these calculations has a minor impact.

If promotion boards use these tiers to make decisions that affect soldiers' careers, women on average would be at a disadvantage when cut points are determined using passers only. Table 3.6 shows the overall proportions of women who fall into each category relative to men using both approaches for establishing cut points.

[6] Both charts in Figure 3.1 contain the entire sample of women for whom we have complete and usable ACFT scores and who are not on profile. Thus, the figure includes women who took the ACFT and passed and those who took the ACFT and failed.

FIGURE 3.1

Proportion of Regular Army Enlisted Women at Each ACFT Fitness Tier, with Cut Points Calculated Two Ways

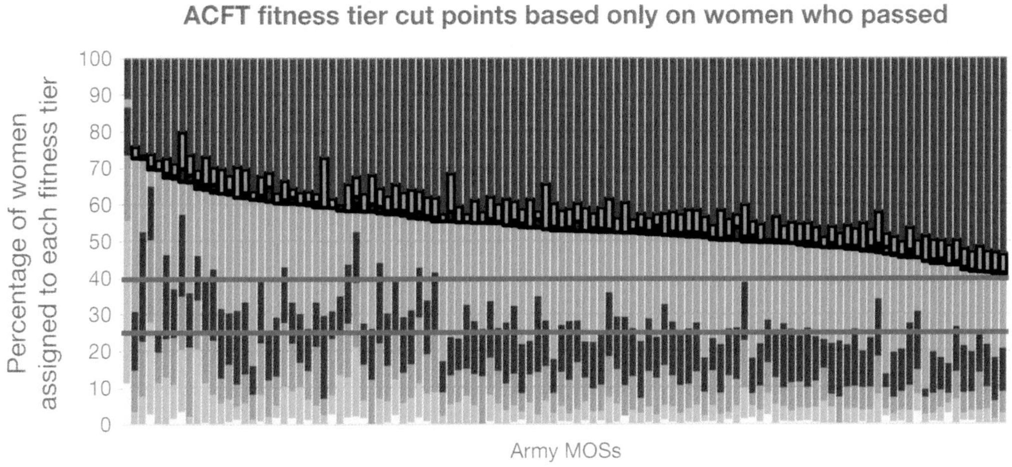

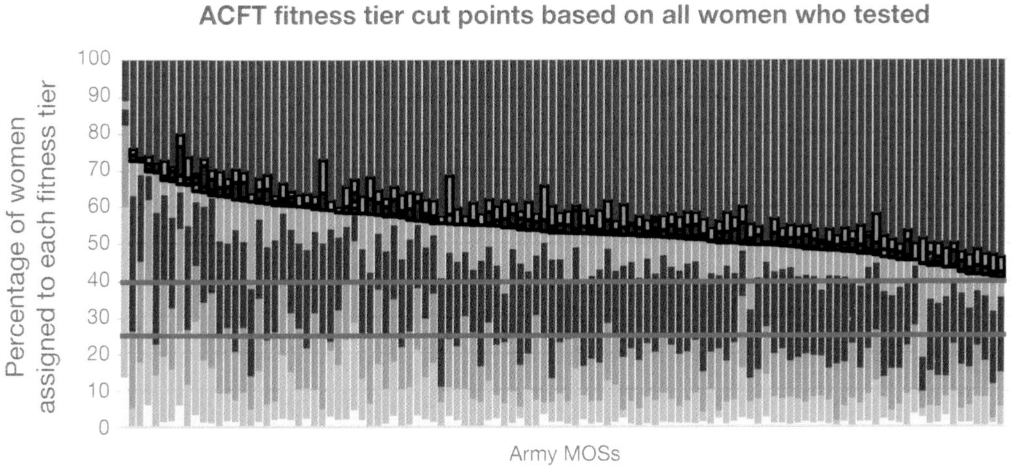

ACFT Tier	Percentile (by gender)	Regular Army Women Overall	Regular Army Women Passing
Platinum	Top 1%	481	498
Gold	Top 10%	422	436
Silver	Top 25%	397	411
Bronze	Top 50%	374	391
Green	360 and up to 50%	360	360
Not Eligible	Below 360		

NOTE: Both figures contain scores from all Regular Army women test-takers. However, tiers (the bottom portion of both figures) are constructed differently according to the two approaches we considered. In the top figure, percentiles are calculated using only passing scores. In the bottom figure, all scores are included when calculating percentiles. Bars outlined in black indicate scores of 360 and above in cases in which soldiers failed the ACFT.

TABLE 3.6
Percentage of Enlisted Regular Army Women in Each Promotion Tier Using Two Approaches to Cut Points

	Percentage in Each Tier When Cut Points Are Based on Passers Only		Percentage in Each Tier When Cut Points Are Based on Entire Population	
Tier	Men	Women	Men	Women
Platinum	0.9%	0.5%	1.0%	1.0%
Gold	8.0%	5.0%	9.0%	9.0%
Silver	14.0%	9.0%	15.0%	14.0%
Bronze	23.0%	14.0%	25.0%	22.0%
Green	46.0%	24.0%	43.0%	7.0%
Failed the ACFT	8.0%	48.0%	8.0%	48.0%

NOTE: Pass rates reported here include the most recent test taken from 2019 through September 10, 2021. Those who completed an alternate event, were listed as being on profile, had missing data on one or more events, or had scores outside the expected range of scores on an event were excluded.

Recommendations to Address Impacts on the Workforce

The Army should consider policies to mitigate group differences in pass rates for at least three reasons. First, if a large portion of the workforce cannot achieve minimum scores on the day the Army institutes the ACFT for record, the Army is making an assertion that those individuals may not be fit for duty. Yet, prior to full implementation of the ACFT, these soldiers may otherwise be in good standing with the Army.

Second, some MOSs will be more heavily affected, which could lead to retention and recruiting issues, and downstream effects on readiness, in those occupations in the future. For nonphysically demanding MOSs, this trade-off of losses from retention and recruiting must be balanced against any gains to physical ability that might be achieved.

Third, although some men are not currently meeting the minimums, the available data suggest that women represent the bulk of the workforce that have the potential to go from "in good standing" to "not in good standing" following the implementation of the test for record (recall from Table 3.1 that, among the enlisted population, pass rates for women range from 41 percent to 52 percent, depending on component, whereas pass rates for men range from 83 percent to 92 percent). As a result, much of the potential impact of the current pass/fail policies will fall on women. The Army therefore needs to decide whether it is willing to accept the impacts that current pass rates will have on the force; if so, the consequences must be fully justified. The consequences for members of other groups, including older soldiers who have also experienced lower pass rates, should be similarly evaluated and justified.

Any adjustments the Army makes to the test or its policies must be aligned with the goal of achieving its three purposes: ensuring physical fitness needed for combat, reducing preventable injuries or ensuring physical health, and transforming Army fitness culture. Policies should not be written for the sake of improving pass rates alone.

Change How the ACFT Is Scored to Mitigate Impacts and Align Requirements with Job-Specific Physical Demands

The Army should consider implementing the following changes to how the ACFT is scored:

- **Revisit event minimums.** The Army could decide to change event minimums for all soldiers (for an example of how pass rates would change under minimums of 40 or 50, see Appendix F). Alternatively, group minimums could be established. For example, for MOSs that do not deploy to combat environments or do not routinely perform combat tasks, the Army could consider lowering minimum scores for passing while holding MOSs with high physical or combat demands to the current standards. Minimums could also be outcome-based: i.e., a specific target for acceptable cardiovascular endurance could be established, and within-gender and within-age minimums could be determined.
- **Norm the ACFT.** The Army made the decision to make the ACFT gender- and age-neutral, meaning raw scores on events (e.g., completing the two-mile run in 19 minutes) translates to the same score for all solders (19 minutes translates to 65 points), regardless of age or gender. This is different than the APFT, which was gender- and age-normed. Norming the test would acknowledge physiological differences between genders,[7] performance differences as soldiers age, and the combat and physical demands of a particular job (if the test were also MOS-normed). Norming the test would ensure parity in pass rates between groups, but it would also require the Army to accept differences in potential combat readiness among soldiers who are held to different testing standards.

The use of lower minimums and group norms could be a temporary solution to allow soldiers to continue to train for and complete the ACFT. Over time, minimums could be increased, and norming could be adjusted or even eliminated. If either approach is used, the scoring tables should be monitored and adjusted as more soldiers take the test and pass rates improve. Recall that only 70 percent of the Regular Army enlisted population, and much lower percentages among other groups, have tested even once. Scoring distributions and pass rates will change as more soldiers test and as more are able to pass.

Review and Update Fitness Policies to Ensure a Mechanism Is Available to Address Exceptional Cases or Circumstances

Establishing policy exceptions is another way to approach mitigating low pass rates among certain groups or for individuals. Under the APFT, several exceptions were put in place for soldiers with certain characteristics, such as returning from deployment and being in certain age groups, and for some occupations. For example, judge advocates and Army Medical Department officers who incurred service obligations because of education and training par-

[7] See, for example, Bishop, Cureton, and Collins, 1987; and Courtright et al., 2013.

ticipation could not be discharged for two consecutive failures on the APFT, as was the policy for other officers (Army Regulation 600-8-24, 2020). The Army should consider continuing these policies, even temporarily, under the ACFT to mitigate the impacts for those and other groups who fail at higher rates.

Use Data from All Test-Takers to Establish Fitness Tier Cut Points

The two approaches to establishing fitness tiers—using only passing scores or using all scores—result in stark differences for women. If the intent of the fitness tiers is to account for physiological differences between men and women in a way that allows the Army to reward and recognize excellence among all soldiers equally (meaning that similar proportions of each gender would be represented in each tier), we recommend norming based on all test-takers in each gender, not only on those who pass the test. Doing so would also produce more-stable tiers: When using passers only, the number of soldiers eligible to be assigned to a tier, and the number in each tier, is likely to change more over time.

Collect and Analyze Data on the Impacts on the Workforce, Including Recruiting, Retention, Promotion, and Unit Readiness

The Army asked us to examine the effect of the ACFT on recruiting, retention, promotion, and unit readiness outcomes. Because test scores during the diagnostic phase are not being used to take administrative actions or for personnel decisions, any attempt to describe these relationships would be speculative. We recommend that the Army collect and analyze data on the relationship between test performance and career outcomes and readiness. These analyses should include linking ACFT and other personnel data to analyze relationships, accurately documenting when a decision (e.g., being denied a promotion) was made because of a soldier's performance on the ACFT, and collecting data from prospective recruits about why they did not join the Army and from soldiers about retention decisions, including whether the ACFT was a factor in their decision.

CHAPTER FOUR

Evidence for Training and Implementation Alternatives to Increase ACFT 3.0 Pass Rates

In addition to the policy levers described in the previous chapter, evidence suggests that access to training can be important in improving scores. Thus, the Army might be able to reduce some of the ACFT pass rate gaps by making investments in training and other supports. In this chapter, we describe findings from three analyses that examined the relationship between training and ACFT outcomes, followed by a summary of themes on training that we heard during discussions and focus groups.

Data Show That Soldiers Do Better When They Train or Have the Opportunity to Retest

The available diagnostic data offer limited information about score improvement because of the lack of available data on pre-ACFT training levels and because most individuals who have tested have done so only once.[1] However, we examined the subset of soldiers who had tested twice in Basic Combat Training (BCT). In addition, we identified two other studies that explored improvements in test scores in different contexts: (1) ACFT scores for U.S. Military Academy (USMA) cadets and (2) ACFT scores after soldiers changed their training regimen in preparation for the ACFT.

Basic Combat Training

During BCT, individuals are fully immersed in a physically demanding training environment and are highly motivated.[2] Therefore, training during BCT happens in an environment with incentives closer to the operational environment envisioned after the implementation of

[1] As a reminder, only a subset of the workforce had completed the ACFT as of September 10, 2021—about 70 percent of Regular Army enlisted soldiers. For officers and the other components, the proportion with at least one test on file was much lower. See Appendix A for details.

[2] An explicit goal of BCT is to build up the physical skills and abilities of new soldiers, and a good portion of BCT activities are dedicated to this goal.

the ACFT and can provide some evidence about the likely effects of training in that environment. First and last scores on individual ACFT events for men and women who tested during BCT indicate notable improvements in pass rates between the two tests, suggesting that pass rates can improve with training (Figure 4.1).

Because of the nature of this sample, there are limits to how much these results can be generalized. It is likely that overall improvements made by the broader workforce will follow different patterns than that of newly enlisted soldiers. The rest of the workforce is necessarily older and, in most cases, cannot dedicate a similar amount of their day to training for the ACFT. Nevertheless, the results suggest that, with enough time and the right training, pass rates for women in particular could measurably improve. Although improvement can and should be expected to continue, it will likely plateau at some point. It is not yet clear whether that improvement will fully eliminate the difference in pass rates between men and women or simply reduce it. In addition, we would always expect differences in the average scores of men and women on some of these events, because physical ability differences across the genders in certain areas of physical performance are well-documented and have been shown to persist even among elite athletes.[3]

FIGURE 4.1
Improvement in Event Pass Rates During Basic Combat Training, by Gender

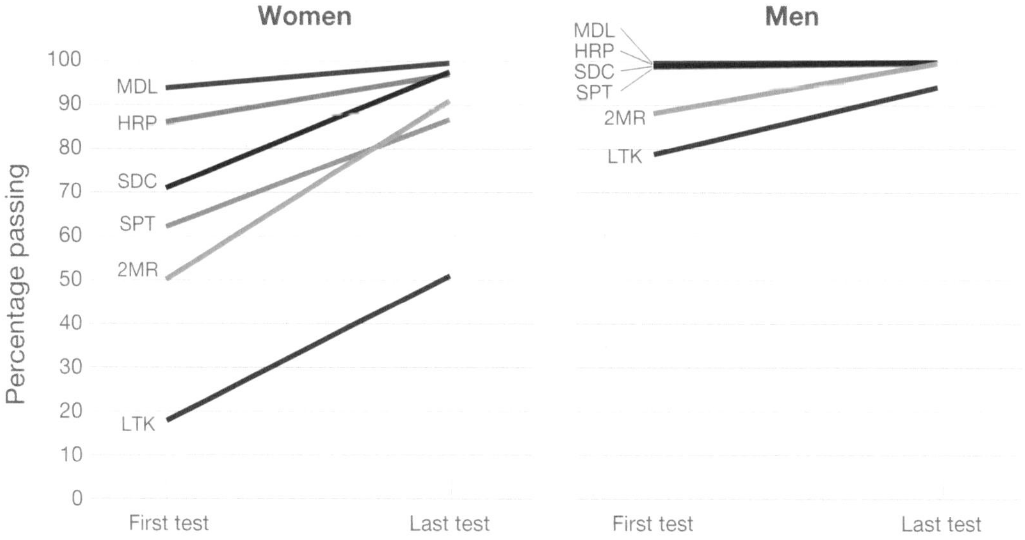

NOTE: MDL = maximum deadlift; SPT = standing power throw; HRP = hand release push-up; SDC = sprint-drag-carry; 2MR = two-mile run; LTK = leg tuck. Sample sizes are 1,167 for women and 5,345 for all events shown. The plank was not included because the sample size in BCT was too small. Per our conversations with U.S. Training and Doctrine Command leadership, the plank was not permitted in BCT during the diagnostic period. Instead, all BCT trainees were required to complete the leg tuck. For this reason, we include in these pass rates people whose leg tuck values were zero even when the plank score was also listed as zero.

---

[3] For example, see Bishop, Cureton, and Collins, 1987; and Courtright et al., 2013.

U.S. Military Academy Data

The USMA conducts ACFT testing at least twice per year. Table 4.1 provides the average results for these two test administrations by class graduation year. USMA attributed these sizable improvements in ACFT pass rates across the classes to greater test familiarity, focused training, and implementation of the plank event as an alternative to the leg tuck.[4] We also note the performance differences by class graduation year: Cadets with more seniority passed at higher rates for both men and women.

Exposure to Different Amounts of Training

According to the *2019 Physical Training and Readiness Survey Report*, initial indications are that the diversity of ACFT events resulted in modifications to personal training efforts: 83 percent of responding soldiers in 2019 were changing their physical training routine to prepare for the ACFT (Muraca-Grabowski, 2020). This is generally consistent with the intended changes to fitness culture that the Army is hoping to see.

After the introduction of the ACFT, the Army modified the training options that were available to unit commanders to complement the new ACFT fitness events and offered revised training options to better prepare soldiers for the battery's introduction.[5] U.S. Army Public Health Command conducted analyses to explore whether there were changes in people's personal training and in the unit training they were receiving after the ACFT was introduced.

TABLE 4.1

Improvement in U.S. Military Academy Cadets' ACFT Performance on Second Test Administration

Class Graduation Year	Average ACFT Score		Change in ACFT Scores
	Spring 2020 Administration	Fall 2020 Administration	
2021	495	514	+19
2022	490	514	+24
2023	468	491	+23

SOURCE: Summary statistics provided by the Army in response to RAND request for information.

NOTE: The 2024 class had yet to matriculate by spring 2020. However, according to its fall 2020 results, 92 percent of cadets passed after four weeks of Cadet Basic Training (shortened because of the COVID-19 pandemic); 99 percent of men passed, and 83 percent of women passed.

[4] The addition of the plank reflects an important limitation to conclusions that can be drawn from the USMA results. It is unclear how many cadets failed the leg tuck initially and how many chose to test on the plank during the second administration. We also do not know how much of the point increase was due solely to that change. No details on this were provided in the briefing shared with us.

[5] For example, they can now use kettle bells, ammo cans, and water bottles to prepare for the ACFT, whereas before the focus was on preparing for sit-ups, push-ups, and running. See Field Manual 7-22, 2020.

U.S. Army Public Health Command also explored whether that training affected ACFT outcomes for both men and women.

Table 4.2 compares the outcomes for each ACFT fitness event for different samples of soldiers who were exposed to the revised unit training program for varying amounts of time prior to taking the ACFT. The overall findings show that both men and women significantly improved their total ACFT scores using the revised training regimen (which placed greater emphasis on resistance training and deemphasized running). Accordingly, men showed a general progression of improvements for all events except the two-mile run. Similar significant results or favorable trends were observed for women, with two exceptions: no improvements for the two-mile run or the standing power throw.[6] It is not known from these analyses whether the levels of performance improvements have plateaued, whether there is further growth potential, or whether these levels of achieved performance can be sustained over longer periods. Nevertheless, these results suggest that changes to unit fitness programs have the potential to improve ACFT outcomes. The lack of improvement for the two-mile run by both men and women was consistent with the reduction in running as part of soldiers' training.

In addition, some studies have reported benefits from establishing tailored training programs in military settings and have suggested that these benefits may be especially important for supporting success among women.[7]

Soldiers Want More Access to Equipment and Tailored Training and Coaching

In addition to our analysis of diagnostic ACFT data, we held conversations with leaders, physical fitness training experts (Holistic Health and Fitness [H2F] personnel, master fitness trainers [MFTs],[8] physical therapists, and physical fitness trainers), and members of the work-

---

6 Women did not experience statistically significant improvements on the maximum deadlift or the leg tuck, either. However, in terms of magnitude, they made meaningful gains.

7 See, for example, Courtright et al., 2013; Dada et al., 2017; Kraemer et al., 2001; Nindl, 2015; and Varley-Campbell et al., 2018.

8 *Master fitness trainers* are personnel who attend the master fitness trainer course. According to the Army,

> The Master Fitness Trainer Course trains selected Noncommissioned Officers (NCOs) and Commissioned Officers in all aspects of the Army's Physical Readiness Training (PRT) System, so they can be unit advisors on physical readiness issues and monitor the unit and individual physical readiness program. The MFTC [Master Fitness Trainer Course] is now taught in a two-phase format. The first Phase consists of 60 academic hours of Distributive Learning (dL) comprised of exercise science classes. The second Phase is a 2 week, 76 academic hour resident course which covers all PRT exercises and drills. You will be expected to tie in the knowledge gained from the DL portion of the course and put it into practice. The second Phase is offered as a resident course at Fort Benning, GA or as an MTT. (U.S. Army Fort Benning and the Maneuver Center of Excellence, 2021)

TABLE 4.2

ACFT Event Performance, by Days Spent Performing Revised Unit Training Before Taking the ACFT

Fitness Event	0–30 days	31–90 days	91–180 days	181 or more days
Average Fitness Event Score for Men				
Maximum deadlift (pounds)	**236**	**246**	**244**	**260**
Standing power throw (meters)	**9.2**	**9.4**	**9.5**	**9.7**
Hand release push-up (reps)	**35**	**37**	**36**	**38**
Sprint-drag-carry (minutes)	**1.92**	**1.87**	**1.88**	**1.84**
Leg tuck (reps)	**7.7**	**8.4**	**8**	**9.3**
Two-mile run (minutes)	16.9	16.7	16.8	16.6
Total score (points)	**457**	**467**	**463**	**476**
Range of sample sizes across events	426–461	136–146	232–249	381–420
Average Fitness Event Score for Women				
Maximum deadlift (pounds)	165	181	171	179
Standing power throw (meters)	5.7	5.8	5.9	6
Hand release push-up (reps)	**24**	**31**	**27**	**28**
Sprint-drag-carry (minutes)	**2.44**	**2.27**	**2.3**	**2.31**
Leg tuck (reps)	1.8	3.4	2.6	3.3
Two-mile run (minutes)	18.3	17.3	18	17.8
Total score (points)	**332**	**374**	**361**	**375**
Range of sample sizes across events	90–100	23–25	42–44	46–52

SOURCE: Grier et al., 2021, pp. 35–36.

NOTE: ACFT fitness event scores were extracted from the digital training management system. Bolded rows reflect a statistically significant improvement in the event metric using analysis of variance tests. Unit training programs were revised in conjunction with ACFT implementation on October 1, 2020.

force. Some of those conversations were informal; others occurred as part of more-formal focus group discussions held with members of the workforce. Formal discussions included conversations with personnel from multiple U.S. Army Forces Command units, ARNG units in two states, and USAR units in multiple locations.[9] Most of these formal groups included both enlisted personnel and officers. Themes from our discussions and focus groups reflect the opinions of only a subset of the force and therefore might not generalize to the entire soldier population.

Force-Wide Findings

Several themes emerged from discussions and focus groups that were common to all components. First, participants discussed the limitations of unit physical training relative to the benefits of individual, tailored training. Unit training often focuses on running or other exercises that are easy to perform in a group, but such exercises may only be needed by a subset of the group and can be done individually and without equipment. Even when a unit focuses on strength training, it can be challenging and time-consuming to set up equipment to make the training useful to soldiers who need to train at very different levels (e.g., on average, women lift less weight than men).

In contrast, multiple participants explained that the Army's H2F program has been a great addition that reflects the Army's attempt to move away from its long-standing approach of applying the same training to everyone (e.g., daily group physical training) to a program that is much more individualized and tailored to each person's specific level and training needs.[10] However, participants repeatedly pointed out that there are not enough H2F personnel to consistently provide tailored training for each unit, let alone the individualized coaching that they believe is needed. It is worth noting, however, that H2F was still being rolled out at the time of these discussions, and the Army is continuing to increase the number of H2F personnel.

Some participants also expressed concerns that MFTs do not have sufficient expertise to provide coaching that is tailored to individual needs regarding how to improve on the ACFT events and prevent injuries, especially with respect to individuals at the lower end of the performance distribution. The MFT course is a one-time course, so many of the MFTs took the course before the ACFT was introduced. There are two ACFT qualification courses (one is

---

[9] Four units in U.S. Army Forces Command participated (1st Armored Division, 4th Infantry Division, 10th Mountain Division, and 101st Airborne Division) for a total of 28 discussions. We also held eight discussions with USAR personnel. Those discussions included soldiers from at least 16 states and territories across the United States. We held four discussions with ARNG soldiers located in two states. These discussions typically included three to five participants. In addition, we held one-on-one discussions with two COL-level representatives (one from the reserve and one from the Regular Army). We held a total of 42 discussions.

[10] For more information on H2F, see the *U.S. Army Holistic Health and Fitness Operating Concept* (U.S. Army, 2020) and Field Manual 7-22, 2020. We note that H2F is not designed to help with ACFT training specifically but rather is meant to improve health more generally.

eight hours, and the other is 24 hours). Some participants who took these courses expressed that they felt confident in administering the ACFT but did not feel confident that they had developed the expertise needed to provide sufficient coaching on the ACFT-specific exercises.

Finally, participants in many of our discussion groups expressed concerns about soldiers using improper form while training or pushing (or being pushed) too quickly to improve, both of which can cause injury. These circumstances may be more likely to occur during group training, during which everyone is expected to perform the same number of repetitions and use the same weights.

Because of these concerns, multiple participants expressed a desire for a new occupational specialty of professional physical trainers to coach units and individuals, similar to H2F and MFTs but with deeper expertise on ACFT training and proper form to prevent injury.

Reserve- and Guard-Specific Findings

During our discussions with members of the USAR and ARNG workforce, participants offered suggestions and comments about topics of particular concern for the reserve components. For example, participants explained that USAR and ARNG personnel are often not colocated with an Army base and therefore do not have the same access to training equipment and space as Regular Army personnel. As a result, some reservists were paying out of pocket for home training equipment, some in remote locations have no access to public gyms, and those who have gyms nearby have to pay for memberships out of pocket. Participants also noted that those in small apartments lack the needed training space to practice some of the activities (such as the sprint-drag-carry) at home.

Participants also expressed concerns about medical coverage if personnel are injured while training on their own. That said, some reservists have personal medical coverage through their employer or pay for it out of pocket. Participants also discussed the difficulties they faced in balancing the ACFT testing and training requirements against other training requirements during their one weekend per month of training time. Some suggested that more annual training days could help. Some also mentioned that, at most locations, they could not keep the ACFT testing equipment set up and that transporting it from a storage location, setting it up, breaking it down, and returning it to storage was especially time-consuming and limited their ability to test people or train them with any frequency.

Recommendations to Support Training Improvements over Time

Phase In Implementation to Allow Time for Individuals to Improve Performance on Specific Events

A phased implementation could help mitigate differences in pass rates among groups (for example, women passing at much lower rates than men, soldiers ages 45 and up passing at lower rates than younger age groups) and allow more time for training and improved performance. In addition, because some tests are not as well supported by the evidence base as others, as discussed in Chapter Two, a phased implementation approach could help mitigate the impacts on personnel decisions while the Army continues to collect and analyze data.

The Army could phase in the ACFT in a variety of ways. Instituting a pass "four out of six" policy to start, followed by a "five out of six" policy and, eventually, a "six out of six" policy, is one example. To illustrate the impact of this approach, Table 4.3 shows that pass rates improve for both men and women under four out of six and five out of six policy alternatives, but rates improve far more for women (though 7 percent of women would still be failing under a four out of six policy). Such an approach would give the force more time to meet the minimum standards while holding soldiers accountable along the way. The recommendations we presented in the previous chapter (e.g., norming the test, lowering minimums) could also be implemented initially but phased out as more soldiers train for, take, and pass the test.

Implement Training Bands to Help Soldiers Train for the ACFT and Increase Pass Rates

In conjunction with phased implementation, the Army should establish event-specific training bands to help increase pass rates. Within training bands, soldiers would participate in training tailored to their individual needs to help them reach a passing level safely. Their improvement should be monitored and tracked to evaluate progress. Training bands would be especially important for soldiers who are not passing but could also be beneficial to individuals who are passing by providing the support they need to move to a higher level of fitness. This approach helps focus training where it is most needed.

Data on the events that men and women tend to fail could be used to inform training bands. Those who failed the ACFT overall tended to fail one of three events: leg tuck, plank, or two-mile run, as shown in Figure 4.2 (USAR and ARNG results are shown in Appendix E). However, a much larger percentage of women fail the leg tuck and, to a lesser extent, the sprint-drag-carry and the standing power throw, relative to men.

TABLE 4.3
ACFT Pass Rates Under Alternative Pass Rate Policies, by Gender (Enlisted Regular Army)

Pass Rate Policy	Men	Women	Overall
Six out of six	92%	52%	87%
Five out of six	99%	79%	96%
Four out of six	100%	93%	99%

NOTE: Estimates use soldiers' most recent ACFT taken from 2019 through September 10, 2021 (*N* = 252,644 men and 39,702 women). All values were calculated using ACFT 3.0 rules for scoring and pass/fail minimums. Estimates for MOSs with fewer than 30 men or women with an ACFT score are omitted. Individuals who completed an alternate event, were listed as being on profile, or who did not have a valid score for all six events were also omitted.

FIGURE 4.2
Failures by Event for Regular Army Enlisted Personnel Who Failed the ACFT Overall

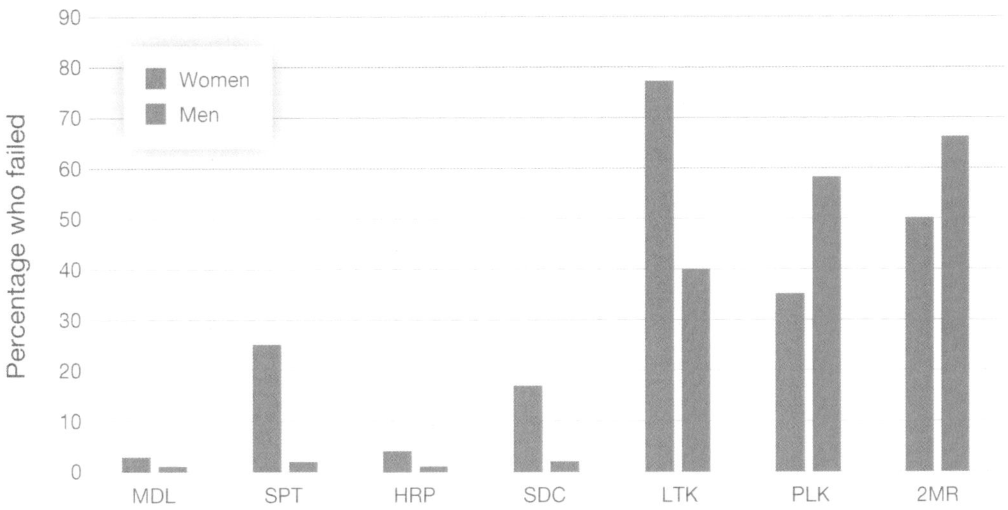

NOTE: MDL = maximum dead lift; SPT = standing power throw; HRP = hand release push-up; SDC = sprint-drag-carry; LTK = leg tuck; PLK = plank; 2MR = two-mile run. Sample sizes are the same for all events except the leg tuck and plank. The number of women overall is 11,018, and the number of men is 14,732. Data for the leg tuck and plank include soldiers who have a score of zero on both events.

Assignment to a training band offers a more supportive approach to those who are failing than the "flag" received under the APFT, which is essentially a derogatory mark on a soldier's record that needs to be corrected before the individual is eligible for a variety of career opportunities. Instead, the Army could approach failures on the fitness test as a training issue: Those who are failing would be identified as *in need of training and assistance* as opposed to soldiers who are *in violation of required physical fitness policies.* This change in tone sends a very different message.[11]

Although we advocate the adoption of training bands to help support a training focus rather than a punitive focus, those who are at low performance levels on an event or who fail multiple events face a greater training burden than those at higher performance levels. That is, soldiers at lower levels of performance would likely be in a training status for longer and require training interventions that are more prolonged relative to their higher-performing peers. These soldiers could be pulled away from job duties or have to commit more of their

[11] The Army could still establish a policy that specifies that if satisfactory progress is not achieved within a training band within a specified period (or if a soldier fails to progress repeatedly and without justification), the soldier could face more-punitive action similar to a flag. However, as described in the fundamentals for evaluating a test in Chapter One, the Army should first ensure that any punitive actions taken are based only on those minimums and events for which it has strong evidence of validity and job relevance for the people who will be affected.

personal time to training than those who are performing at higher levels, which could lead to other negative career consequences, including receiving lower performance ratings from supervisors or being passed over for critical assignments. If the ACFT is implemented as it was fielded during the diagnostic phase (i.e., with all soldiers regardless of gender, age, or MOS being held to the same standards, resulting in differential pass rates), women and older soldiers on average will face heavier training burdens and the corresponding potential downstream consequences.

Ensure Soldiers Have Access to ACFT-Relevant Training, Equipment, and Coaching

An important element of a successful rollout is to ensure that the entire workforce has ample access to the requested training resources. Army analysis indicated that ACFT's introduction and associated changes in individual and physical training regimens have resulted in higher near-term injury rates during the diagnostic period (Grier et al., 2021).[12] To help minimize such injuries and improve the training environment in support of the ACFT going forward, we recommend that the Army

- tailor training and equipment access to meet individual needs (training bands could help with this)
- increase individualized access to expert trainers and coaches (e.g., exercise physiologists and physical therapists)
- target resources toward groups that are especially at risk (e.g., those with low pass rates or high injury rates), if resources are constrained
- provide better support and guidance on the prevention of injury while soldiers train for the ACFT (for example, support could include one-on-one coaching and individualized feedback; guidance could include training on proper lifting mechanics, posture, and progression steps)
- continue to monitor training and testing injuries in real time to establish root causes, evaluate and implement intervention programs, and determine the persistence of injury occurrences
- allow more time and provide additional resources to help further instruct soldiers in the proper conduct of and training needed to successfully transition to the full-fitness dimensions of the ACFT.

[12] There have been no systematic force-wide changes to unit training; however, individual units have discretion over the types of physical training that they do, and some units have changed their approach to physical training in a variety of ways. Individuals are also training differently on their own, as discussed earlier in this chapter.

CHAPTER FIVE

Conclusions and Recommendations

In previous chapters, we described the results of our assessment of the ACFT, which included an evaluation of prior work conducted by the Army, an analysis of diagnostic ACFT data showing the potential impact of the test on personnel outcomes, and discussions and focus groups with leaders, practitioners, and members of the workforce. In this chapter, we conclude with key findings and recommendations, which fall into four broad areas: the evidence base, impacts on the workforce, training, and the need for a formal management structure. The recommendations, summarized in Box 5.1 and discussed in the sections to follow, are aimed at identifying the important actions and considerations that the Army needs to address as the ACFT implementation process continues.

BOX 5.1

Top-Level Recommendations

1. **Address shortfalls in the ACFT evidence base.**
   a. Collect additional data to further explore validity findings by gender and to establish justifiable minimum standards on fitness events.
   b. Establish proper justification for why all ACFT events and minimum standards apply equally to all soldiers.
   c. Continually examine and assess personnel decision outcomes that are associated with minimum scores.
   d. Define and continually assess organizational progress toward fitness transformation and solicit perspectives from the total force.
2. **Consider ways to mitigate impacts on the workforce.**
   a. Change how the ACFT is scored to mitigate impacts and align requirements with job-specific physical demands.
   b. Review and update fitness policies to ensure that a mechanism is available to address exceptional cases or circumstances.
   c. Use data from all test-takers to establish fitness tier cut points.
   d. Collect and analyze data on the impacts on the workforce, including recruiting, retention, promotion, and unit readiness.
3. **Take steps to further support training improvements over time.**
   a. Phase in implementation to allow time for individuals to improve performance on specific events.
   b. Implement training bands to help soldiers train for the ACFT and increase pass rates.
   c. Ensure soldiers have access to ACFT-relevant training, equipment, and coaching.
4. **Institutionalize a formal senior-level management structure to guide and oversee ACFT implementation and use.**

Top-Level Conclusions and Recommendations

The Evidence Base to Support the ACFT Is Incomplete

The Army has demonstrated support for some, but not all, aspects of the ACFT. The most pressing issue for the Army to address is the evidence shortfalls. The Army has gathered a wealth of evidence on the ACFT. But the evidence gathered so far is mixed in its support of some fitness events included in the test, and there are gaps in the evidence base that are important for the Army to fill. As we discussed previously, the leg tuck and plank especially are not well-supported for use in predicting combat task performance or in terms of preventing injuries. These events need additional evidence if the Army plans to use them for record. All events could benefit from additional predictive validity and standard-setting studies on broader and larger samples of both men and women for all three of the ACFT's purposes.

In addition, greater consideration should be given to which soldiers should be held to combat standards versus general health standards. The minimums on the test are focused largely on ensuring minimum levels of combat task performance, and it is not clear that all MOSs or all individuals in any particular MOS need to be held to these standards or that the trade-offs that would result from holding them to those standards are acceptable. Because this test ultimately will be used for personnel actions that can affect the careers of soldiers and because there are many large differences in how women are performing relative to men, the Army must make sure it has strong evidence showing that the events and minimum standards are valid and needed for all who are affected by them.

ACFT Scores Collected During the Diagnostic Period Show Some Groups Failing at Noticeably Higher Rates

Considering the pass rates we observed, it appears that many soldiers would be failing if the test were instituted today, including soldiers who are otherwise in good standing in the Army. The biggest impacts are observed for women, but we also see differences in pass rates across MOSs, age groups, and components, with the USAR and the ARNG lagging behind the Regular Army. Although these differences do not necessarily mean the test is flawed, they need to be investigated further. The Army needs to show that the differences are necessary and unavoidable to directly support mission-critical job requirements for all who are affected by them. The Army also needs to continue to investigate the causes of the differences to identify steps that can be taken to help facilitate improvements in the differences over time. The differences need to be tracked and periodically monitored for changes, and leadership needs to be cognizant of the size of the differences and the resulting effects on the force.

The Army also needs to decide whether these differences are acceptable and be prepared for the consequences (e.g., potential personnel losses and effects on readiness) or to take steps to mitigate any consequences, as described in previous chapters. If scores are factored into personnel decisions or personnel actions, the potential implications for personnel management could be considerable. Because of this potential outcome, the Army may want to put mitigation strategies in place to anticipate and manage areas with the greatest impacts, including MOSs in which overall pass rates are low; specific populations, such as those older than 45, many of whom are in leadership roles; and groups who already are underrepresented in the workforce overall and in certain MOSs, such as women. Approaches to mitigation ideally should be informed by data, including information on the potential trade-offs that might be incurred for those MOSs that are not likely to encounter combat.

Research in Multiple Military Settings Has Shown That Training Can Improve Pass Rates

Training, equipment, and supports are important to set soldiers up for success. The Army has rolled out equipment force-wide, started training soldiers to improve performance on the ACFT, and provided some access to H2F personnel and MFTs. Research has shown that training programs in other contexts can generally have a positive impact on enhancing soldier fitness, but greater effort is needed to better understand and properly design tailored efforts to address distinct Army circumstances, the specific ACFT events, and the individualized training needs of units and soldiers at different starting performance levels. Ensuring training and equipment access for all will need to be a continuous process that is refined and improved over time. The Army should continue to build on current efforts and make improvements to realize the benefits as quickly as possible.

The Army Would Benefit from a Formal Management Structure to Oversee Refinements to the ACFT over Time

The Army must continuously monitor the ACFT after its full-scale implementation. Because operational, force management, and policy considerations related to assessing physical fitness and its force-wide application are interconnected, this role cannot be executed by an existing Army structure or solely monitored by an existing Army organization. For example, a Forces Command element (or Department of the Army Headquarters Operations division), a Training and Doctrine Command entity, or a Department of the Army Headquarters policy component each has distinct equities in the formulation of and outcomes associated with the ACFT and its usage. Similarly, other Army stakeholders have considerable interest in and perspectives about ACFT-related issues, including the reserve components, individual enlisted personnel and officers, the medical community, and individual MOSs. The breadth of interest and organizational equities have hampered past service efforts to consistently implement and oversee physical fitness programs (GAO, 1998).

The Army needs to address critical test and test use issues, expand the test's evidence base, and provide necessary resources and support for soldiers to successfully perform on the test. Because the test is administered for record, the Army will need to continuously monitor the test to ensure it is working as intended. Implementation will raise additional questions regarding

- whether ACFT implementation and the associated personnel decisions are affecting retention, recruiting, health care use, manning in the reserve components, and the availability and deployability of personnel across all specialties
- the best way to monitor these questions and issues going forward
- how the Army will prepare for, address, and resolve these and other issues that will arise and any resulting unintended consequences.

These topics all require cooperation among Army organizations that often have competing objectives and perspectives.

In its development, validation, and eventual full-scale implementation of the ACFT, the Army has acknowledged the need for ongoing and sustained focus by its leadership to ensure that the test is properly executed across the total force and is achieving its stated purposes. The Army has an extensive history associated with physical fitness training, measurement, policy determination, and decisionmaking. In these areas, the Army has demonstrated that it is a learning organization capable of reviewing and revising decisions regarding fitness processes and procedures. That past agility and responsiveness will be essential as the Army faces its most significant changes to its physical fitness test of record in more than 40 years.

Institutional preparation for such fundamental and broad-reaching change to the Army's physical fitness training programs and assessment processes cannot be understated. The involvement and attention of senior leadership in guiding, resourcing, and monitoring all aspects of the process are key (Meredith et al., 2017). Successful institutional change fundamentally requires consistent engagement and communications with soldiers so that they understand the context for the change and the relevance of solutions in addressing the overall challenges. Such change management capabilities do not reside within a single Army organization but rather require coordination and collaboration among multiple entities.

Given the magnitude and complexity of issues involved, we recommend that the Army establish and empower a multiperspective, total-force executive structure to institutionalize ACFT guidance, resourcing, and monitoring as the test is implemented, matures, and evolves. We propose that this executive structure be cochaired by the Under Secretary of the Army and the Sergeant Major of the Army, positions that are of sufficient stature to (1) address the diversity and complexity of issues involved and (2) integrate and effect solutions and resource requirements across the total force. We envision this permanent standing body being composed of general officers and senior executive service representatives who are supported by subordinate working groups, as needed.

The composition of the executive structure should include a broad range of functional areas, including (but not limited to) enlisted and officer personnel and force management policy, military readiness and operational perspectives, individual- and unit-level training, USAR and ARNG considerations, and military medicine. The charter for the executive structure should initially concentrate on implementing guidance, managing change, and communicating; determining, timing, and tracking the effects of resource investments and any additional requirements; conducting additional ACFT supporting analysis; and monitoring and responding to initial testing and personnel outcomes across the force. The standing structure should receive proper analytical support so that its processes and decisions are informed by robust diagnostic designs and data collection.

A Note of Caution About the Data Presented in This Report

We found several problematic errors in data recording and omissions in the data that affect the conclusions that can be drawn from the diagnostic data. For example, we were unable to identify everyone who was on a medical profile and who therefore were unable to complete all or parts of the test.[1] Instead, we removed all individuals who did not have a complete ACFT from our analyses, even if they were not listed as being on a medical profile. However, doing so results in a higher pass rate than if those individuals are included. In addition, we were unable to tell, in many cases, whether someone completed the leg tuck, plank, or both when they had scores of zero for both. If those individuals are included in the estimates for both events, the pass rates are necessarily lower for both. We also found data entry errors (e.g., 60 leg tucks likely should have been one leg tuck, which equates to 60 points) and suspect that there are many similar errors on the other events that are harder to detect because they are within the plausible raw score range.[2]

Although some of these data problems may resolve or self-correct once the ACFT is implemented and soldiers can see their scores and report errors, others will require explicit interventions by the Army to correct. Taking steps to improve data problems is crucial because the Army cannot know how well the ACFT is working unless data are accurate, timely, and complete.

Therefore, the Army should consider the following actions to improve ACFT data-recording practices:

- establish additional data auditing processes during data entry to ensure event scores are being recorded accurately
- periodically check the accuracy of Digital Training Management System calculations (event score values, pass/fail values, and ACFT total scores) to resolve any errors

[1] Medical profiles indicate a soldier's physical limitations associated with a medical condition. Profiles can be temporary or permanent and specify which activities the soldier cannot perform (including fitness test events) and suitable alternatives, if appropriate.

[2] For example, a deadlift of 60 pounds would be a failing score on the maximum deadlift, but if the person actually lifted 140 pounds (which corresponds to a score of 60), then they would receive a passing score; if a 60 is entered in the database, we do not know whether that accurately reflects their raw score or whether it was a mistake.

- electronically link the eProfile system to the ACFT system (no hand entry) and ensure that everyone on a profile appears in the data
- for optional events (currently, the leg tuck and plank), ensure the data system indicates which event was attempted
- incorporate temporary profiles into the record system.

Closing Thoughts

This report offers a summary of our independent review. We have noted strengths and weaknesses of the ACFT development and validation and presented a variety of specific recommendations to support the test's successful implementation. Some of these recommendations are to change policies (e.g., adjusting the way the test is scored, establishing exceptions for how the test is used for some soldiers or groups of soldiers), either temporarily or permanently, to mitigate impacts on the workforce.

However, these recommendations require the Army to grapple with a trade-off: Should all soldiers, regardless of age, gender, and MOS, be subject to the same physical fitness standards (which prioritize combat readiness), or should the test and policies be set such that all soldiers have a reasonable chance of passing but not at the same minimum level of physical fitness? Regardless of how or whether the Army fully implements the ACFT in April 2022, central to our recommendations is the Army's commitment to continued senior leadership oversight and guidance in monitoring the ACFT's implementation over the long term and to making data-supported refinements to ensure that individual soldier interests and operational equities are adequately considered and balanced.

APPENDIX A

Diagnostic ACFT Data Sample Sizes and Proportion Tested, by Component

This appendix contains sample sizes of the soldiers who took the ACFT during the diagnostic period and the portion of the workforce reflected. Table A.1 shows the approximate proportion of officers and enlisted personnel in each component that had at least one ACFT record in the Digital Training Management System as of September 10, 2021.

TABLE A.1

Number and Proportion of Soldiers with at Least One ACFT Score, by Component, Enlisted or Officer Status, and Gender

Component	Enlisted or Officer	Gender	Number of Soldiers with at Least One ACFT Score	Approximate Proportion of the Workforce That Has Tested
Regular Army	Enlisted	Women	35,394	49%
		Men	244,058	71%
		Overall	279,452	70%
	Officer	Women	7,852	52%
		Men	33,436	60%
		Overall	41,288	58%
USAR	Enlisted	Women	8,109	23%
		Men	28,967	30%
		Overall	37,076	28%
	Officer	Women	995	15%
		Men	4,097	23%
		Overall	5,092	21%
ARNG	Enlisted	Women	15,818	39%
		Men	73,803	45%
		Overall	89,621	44%
	Officer	Women	1,364	34%
		Men	9,034	45%
		Overall	10,398	43%

NOTE: The proportion of the workforce is approximate and based on a single snapshot of the workforce size, as of August 2021.

Details on Improving Validation of the ACFT

The Army conducted the BSPRRS to

- "determine the baseline physical readiness requirements of the physically demanding, commonly occurring and critical Warrior Tasks and Battle Drills and Common Soldier Tasks (WTBD/CST)"
- "determine if the current 3-event Army Physical Fitness Test (APFT) adequately assessed the baseline physical readiness required to accomplish physically demanding WTBD/CSTs"
- "determine if there were other physical fitness test events that better predicted Soldier performance on physically demanding WTBD/CSTs" (East, DeGroot, and Muraca-Grabowski, 2019, p. iii).

The final outcome of the study was a physical fitness test (the ACFT) that allowed the Army to predict WTBD/CST performance. The authors conclude that the ACFT "provides acceptable predictive validity ($R^2 > 0.835$, $p = 0.000$) to identify Soldiers capable of executing high-demand, commonly occurring and critical WTBD/CSTs" (East, DeGroot, and Muraca-Grabowski, 2019, p. iii).

Throughout this report, we highlighted weaknesses in the BSPRRS's finding that the ACFT (or individual events) can predict performance on combat tasks (and we highlighted the need for additional research to support injury prevention and a change in fitness culture). We have focused our attention mostly on concerns about the leg tuck and plank. In this appendix, we provide additional detail about our concerns. As the Army considers its implementation of the ACFT's six events and alternate event, or as it considers modifying the test or individual events in the future, the principles we describe here should guide those decisions.

A main limitation of the BSPRRS is sample size. The effort to validate a fitness battery (confirm that performance on the test predicted combat tasks) took place at two locations: Fort Riley and Fort Benning. At Fort Riley, 46 women and 278 men participated in the testing. At Fort Benning, 16 women and 136 men participated. In other words, the performance of 62 women and 414 men was used to determine whether a set of fitness events predicted outcomes on the WTBD-ST. Findings for some individual events (two-mile run, hand release push-up) did not predict WTBD-ST performance for women, despite support in the literature (see Table 2.1).[1] Sample size could be one explanation for anomalous findings in the BSPRRS. In other words, it is possible that differences in validity by gender are occurring by chance alone. In addition, the BSPRRS reflected a narrow range of performance (in terms of both combat and fitness measures) among women, which could have affected the results.[2] To bolster support for individual events and for the ACFT as a whole, the Army should collect and analyze data from larger and more diverse samples of soldiers that cover the full performance continuum.

The outcome studied in the BSPRRS was total time to complete a series of WTBD simulation events. However, it is possible that one WTBD-ST activity dominated the time it took to complete the simulation, and therefore one event could have driven the relationships observed in the study.[3] If performance on each of the WTBD-ST activities was examined and predicted separately, stronger relationships for the different candidate events might have been observed. Using a single composite outcome measure might explain why the two-mile run did not predict WTBD completion times for women. The time to complete the two-mile run may successfully predict performance on a cardiovascular-heavy WTBD task; however, if that task was combined with several others that test different fitness components, the relationship may have been masked. In addition to examining WTBD-ST as an outcome, the Army should explore the relationship between individual fitness events and each of the five individual combat tasks comprising the WTBD-ST.

It is also important to understand and assess what a fitness event is actually measuring. To explain, we will use the stated purposes for the leg tuck, which are two-fold: (1) to measure core strength and endurance and (2) to predict whether a soldier can complete a specific combat task, such as scaling a wall. If the leg tuck is used to measure core strength and endurance, the Army needs to demonstrate that the leg tuck is in fact operating as a measure

1 Whereas the two-mile run and hand release push-up are supported by the broader literature, the leg tuck is a newer event that has not been studied extensively. Therefore, support for it, or the lack thereof, can only come from the Army's work to study it. Although the BSPRRS found evidence of predictive validity for the sprint-drag-carry, it is a multifitness component event that was created for the ACFT and has not been studied specifically in the broader literature. Given the small sample size used to validate the sprint-drag-carry (which was only conducted at Fort Benning and involved 16 women and 136 men), the event should be validated on a much larger sample. The plank, on the other hand, has been studied (though to a lesser extent than some events, such as the sit-up), but the BSPRRS did not include this event, and therefore the Army lacks evidence on its validity in predicting WTBD-ST performance. Because all soldiers self-select whether to perform the leg tuck or the plank, data collected to date do not involve a properly designed study to adequately assess the validity of the plank.

2 There are statistical corrections that can be applied to data to estimate the relationship that might be observed if a larger range of performances were included. However, we would not advise relying on those corrections in this case, given the small sample size and the possibility of unstable estimates for the correlations that would be used to make the correction.

3 Using data from the BSPRRS, it may be possible to explore the relationship between ACFT event performance and how long it took soldiers to complete individual WTBD activities. However, some of the other limitations of the study would still apply (a small sample of women and a narrow range of performance by women on both the predictor and outcome variables), so results would have to be interpreted cautiously.

of core strength and endurance for all individuals. Analyses conducted for the Army by the University of Iowa have shown that the leg tuck requires a combination of upper body and lower body strength and endurance (see Abdel-Malek et al., 2021). As a result, if a soldier has insufficient upper body strength to pull their body weight up toward the bar during a leg tuck, then the strength of their core would not matter. For those soldiers, the fitness event is not measuring their core strength; it would only be measuring their (lack of) upper body strength. No research has been conducted on whether the lack of upper body strength is masking the ability to accurately measure core strength for some soldiers. This could be explored by administering leg tucks and other measures of core strength that do not involve upper body strength to a sample of men and women (for example, the Army has incorporated the plank into the ACFT for the purpose of testing core strength, so it could be used). If it turns out that performing a leg tuck is not correlated with those other measures of core strength and endurance, especially for individuals with low upper body strength, then the test may not actually be measuring core strength.

Second, if the purpose of the leg tuck as an ACFT event is to test the ability to scale a wall, it should be true that soldiers who cannot complete one leg tuck also cannot scale a wall. This has not been tested empirically. As noted earlier, the best way to assess performance on a task is to measure the task directly.[4] However, a leg tuck is not a direct measure of someone's skill at wall-climbing. Instead, it is a more indirect measure that is intended to simulate and capture some of the types of core and upper body strength movements that might help when climbing a wall.

To explain further, the ability of a test (or event or simulation) to serve as an assessment of someone's performance on specific job tasks depends, in part, on the test's fidelity to those job tasks. High-fidelity tests are very close in form and appearance to the tasks that are being tested; low-fidelity events are not close in form or appearance to the tasks they are intended to predict. The lower the fidelity of an event to the actual job, the more analytical support (in the form of predictive validity) is needed to demonstrate and verify the link between perfor-

[4] The WTBD-STs that were used in the BSPRSS are examples of work sample simulations that are intended to replicate WTBD task performance. These simulations are different from the predictors of that performance that were included in the ACFT. That is, the WTBD-STs can be thought of as a proxy for WTBD performance and other combat task performance. The ACFT is not intended as a proxy for that performance. Instead, it is better thought of as a measure of the physical abilities that can help people succeed at those combat tasks (i.e., it is a predictor of the outcome of interest but not a direct measure of the outcome of interest). We are drawing this distinction between predictors of the outcomes and measures of the outcomes; in reality, however, there is not a clear line that separates what would be considered a simulation of a job task versus a measure of the underlying abilities needed to perform the task. Measures exist along a continuum of fidelity to the outcome: Those that have very high fidelity to the work tasks can be thought of as proxies of the work tasks, and those that have little fidelity to the work tasks should not be considered proxies of the work tasks. As noted previously, some of the ACFT events show more-obvious similarities to the combat tasks (the sprint-drag-carry, in particular), but most are more separated from the actual combat tasks.

mance on the event and performance on the job.[5] In the context of combat tasks, high-fidelity physically demanding work samples might entail dragging a dummy out of a vehicle, carrying sandbags and stacking them to build a wall, or scaling a mud brick wall, for example.

Some of the events on the ACFT have more fidelity to combat tasks than others. For example, the two-mile run and the standing power throw are not designed to approximate combat tasks, whereas the sprint-drag-carry approximates such tasks as extracting a casualty on the battlefield. However, the ACFT is not intended as a direct work sample measure of WTBDs, combat tasks, or other physically demanding soldier tasks. That is, the events should not be thought of as direct proxy measures of these activities. Instead, the ACFT should be thought of as a set of measures of physical abilities that are needed to perform tasks, even for events that look similar to tasks performed in combat. The events were designed and selected to assess important physical components that were hypothesized to be relevant and necessary to perform well on WTBDs, combat tasks, or other physically demanding tasks. Assuming that the fitness events are valid predictors of performance, they can also serve as a guide for soldiers in terms of the underlying physical abilities they should develop to have the prerequisite physical abilities to support success on those WTBDs. However, because the ACFT events are not direct measures of these tasks, it is critical to demonstrate a strong linkage between the event and those tasks.

One final note about the fact that ACFT events are not actual measures of whether someone can perform WTBDs and other combat tasks: WTBDs and combat tasks require not only physical abilities but also techniques and other skills that may have to be learned and practiced. The best assessment of how well someone can scale a wall will ultimately be to scale a wall. It may be that some people who can perform multiple leg tucks will not have learned or practiced the techniques for performing a wall climb. Those techniques might only be developed through the practice of scaling walls. Therefore, Army leaders should not assume that performance on the ACFT will automatically translate to acceptable performance on WTBDs without additional practice and training on those WTBDs.

---

5 For a broader discussion of this issue, see Callinan and Robertson, 2000; and Motowidlo, Dunnette, and Carter, 1990.

Physical Tasks Required of All Soldiers in Combat and a Resulting List of Fitness Event Predictors

The genesis for the Army's efforts to define and measure the physical tasks that are required of all soldiers in combat began with the BSPRRS. This research was started in 2012 by a consortium of Army fitness-relevant organizations: CIMT, the U.S. Army Research Institute of Environmental Medicine, U.S. Army Public Health Command, and the USMA. U.S. Training and Doctrine Command chartered the study to "determine the baseline physical readiness requirements of the physically demanding, commonly occurring, and critical Warrior Tasks and Battle Drills" (East, DeGroot, and Muraca-Grabowski, 2019, p. 12).

The initial phase of the BSPRRS involved efforts "to identify physical performance demands and thresholds across a range of representative Soldier types" (East, DeGroot, and Muraca-Grabowski, 2019, p. 11). That is, the Army approached the job analysis stage in the BSPRRS by creating a detailed listing of combat-related tasks expected of all soldiers. The research team considered tasks that are expected to be performed by all soldiers, regardless of MOS, as specified in Army manuals (Soldier Training Publication 21-1-SMCT, 2019). Army policy is that units routinely train to and assess individual performance on a subset of these tasks that are determined to be consistent with the unit's overall mission and the commander's areas of emphasis. Figure C.1 shows the Army's job analysis process for identifying physical combat tasks.

The Army's current list of warrior tasks is shown in Table C.1, with physically demanding warrior tasks (according to East, DeGroot, and Muraca-Grabowski, 2019) in bold, followed by the current list of battle drills. According to the Army's *Soldier's Manual of Common Tasks: Warrior Skills Level 1*, *warrior tasks* are "a collection of individual Soldier skills known to be critical to Soldier survival," such as weapons training, tactical communications, urban operations, and first aid, and *battle drills* are "group exercises designed to teach a unit to react and survive in common combat situations" (Soldier Training Publication 21-1-SMCT, 2019, p. 1-1).

FIGURE C.1

Army Job Analysis Conducted to Identify Relevant Physical Combat Tasks

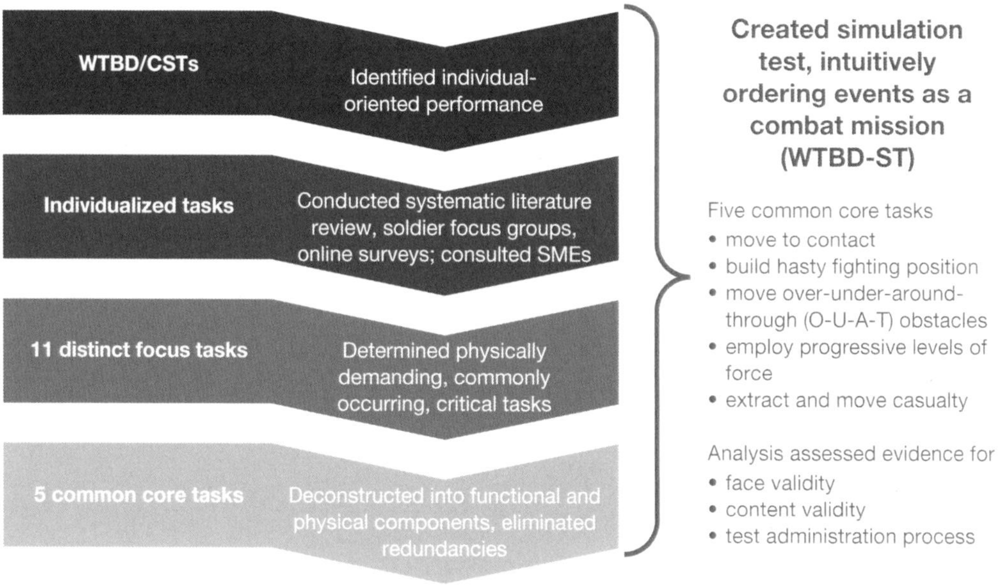

SOURCE: Analysis of East, DeGroot, and Muraca-Grabowski, 2019.

TABLE C.1
Warrior Tasks

Task Number	Task Description
071-COM-0032	Maintain an M16- series Rifle/M4 series Rifle Carbine
071-COM-0029	Perform a Function Check on an M16- Series Rifle/ M4 Series Carbine
071-COM-0028	Load an M16- Series Rifle/M4 Series Carbine
071-COM-0027	Unload an M16 Series Rifle/M4 Series Carbine
071-COM-0030	Engage Targets with an M16-Series Rifle/M4 Series Carbine
071-COM-0033	Correct Malfunctions of an M16- Series Rifle/M4 Series Carbine
071-COM-0031	Zero an M16- Series Rifle/M4 Series Carbine
071-COM-4401	Perform Safety Checks on Hand Grenades
071-COM-4407	Employ Hand Grenades
071-COM-0541	**Perform Exterior Movement Techniques during an Urban Operation**
071-COM-0503	**Move Over, Through, or Around Obstacles (Except Minefields)**
071-COM-1000	Identify Topographic Symbols on a Military Map
071-COM-1001	Identify Terrain Features on a Map
071-COM-1008	Measure Distance on a Map
071-COM-1002	Determine the Grid Coordinates of a Point on a Military Map
071-COM-1005	Determine a Location on the Ground by Terrain Association
071-COM-1012	Orient a Map to the Ground by Map-Terrain Association
071-COM-1011	Orient a Map Using a Lensatic Compass
071-COM-1003	Determine a Magnetic Azimuth Using a Lensatic Compass
071-COM-1006	**Navigate from One Point on the Ground to another Point while Dismounted**
071-COM-0501	**Move as a Member of a Team**
071-COM-0502	**Move Under Direct Fire**
071 COM-0510	**React to Indirect Fire while Dismounted**
071-COM-0513	Select Hasty Fighting Positions
113-COM-2070	Operate SINCGARS Single-Channel (SC)
113-COM-1022	Perform Voice Communications
171-COM-4079	Send a Situation Report (SITREP)
171-COM-4080	Send a Spot Report (SPOTREP)
071-COM-0608	Use Visual Signaling Techniques
031-COM-1036	Maintain Your Assigned Protective Mask

Table C.1—Continued

Task Number	Task Description
031-COM-1035	Protect Yourself from Chemical and Biological (CB) Contamination Using Your Assigned Protective Mask
031-COM-1019	React to Chemical or Biological (CB) Hazard/Attack
031-COM-1040	Protect Yourself from CBRN Injury/Contamination with the JSLIST Chemical-Protective Ensemble
031-COM-1013	Decontaminate Yourself and Individual Equipment Using Chemical Decontaminating Kits
031-COM-1037	Detect Chemical Agents Using M8 or M9 Detector Paper
031-COM-1021	Mark CBRN-Contaminated Areas
031-COM-1018	React to Nuclear Hazard/Attack
031-COM-1042	Protect Yourself from CBRN injury/contamination when changing MOPP using JSLIST Chemical Protective Ensemble
081-COM-1044	Perform First Aid for Nerve Agent Injury
081-COM-1001	Evaluate a Casualty (Tactical Combat Casualty Care)
081-COM-1003	Perform First Aid to Clear an Object Stuck in the Throat of a Conscious Casualty
081-COM-1005	Perform First Aid to Prevent or Control Shock
081-COM-1023	Open An Airway
081-COM-1032	Perform First Aid for Bleeding of an Extremity
081-COM-1046	**Transport a Casualty**
081-COM-1007	Perform First Aid for Burns
081-COM-1026	Perform First Aid for an Open Chest Wound
081-COM-0101	Request Medical Evacuation
052-COM-1270	React to Possible Improvised Explosive Device (IED) Attack
052-COM-1271	Identify Visual Indicators of an Implosive Device
071-COM-0804	Perform Surveillance without the Aid of Electronic Device
301-COM-1050	Report Information of Potential Intelligence Value
071-COM-0815	Practice Noise, Light, and Litter Discipline
071-COM-0801	Challenge Persons Entering Your Area
071-COM-0512	**Perform Hand-to-Hand Combat**[a]
071-COM-0006	React-to-Hand-to-Hand Combat (Repeat)
071-COM-4408	Construct Individual Fighting Position
052-COM-1361	Camouflage Yourself and Your Individual Equipment

Table C.1—Continued

Task Number	Task Description
071-COM-0011	Employ Progressive Levels of Individual Force
181-COM-1001	Conduct Operations According to the Law of War
191-COM-0008	Search an Individual in a Tactical Environment
159-COM-2026	Identify Combatant and Non-Combatant Personnel & Hybrid Threats

SOURCE: Soldier Training Publication 21-1-SMCT, 2019.

NOTE: Physically demanding tasks (according to East, DeGroot, and Muraca-Grabowski, 2019) are bolded.

[a] East, DeGroot, and Muraca-Grabowski, 2019, lists 071-COM-0006 ("React to Man-to-Man Contact") as a physically demanding task. According to Soldier Training Publication 21-1-SMCT, 2019, 071-COM-0006 is "React-to-Hand-to-Hand Combat (Repeat)," but the manual does not provide instructions for this task. Therefore, we have marked 071-COM-0512 ("Perform Hand-to-Hand Combat") as the physically demanding task, even though the task number does not match what appears in East, DeGroot, and Muraca-Grabowski, 2019.

The *Soldier's Manual of Common Tasks: Warrior Skills Level 1* also lists five battle drills:

- react to contact
- establish security at the halt
- perform tactical combat casualty care
- react to ambush (near)
- react to ambush (far).

The Army research team narrowed the collective tasks by applying two criteria: tasks that are individually oriented and tasks that soldiers and subject-matter experts consider to be highly representative of a combat environment. The efforts resulted in a list of 11 focus tasks, as follows:

- physically demanding WTBDs
  - move as a member of a team
  - perform exterior movement techniques—urban operation
  - move under direct fire
  - react to indirect fire dismounted
  - move over, under, around, and through obstacles
  - transport a casualty
  - react to man-to-man contact
  - navigate point to point dismounted
- physically demanding CSTs
  - conduct dismounted tactical foot march
  - prepare a fighting position (fill and emplace sandbags)
  - drag a casualty to immediate safety—mounted.

The research team further refined these 11 tasks into five common core tasks that they determined to be physically demanding, commonly occurring, and critical for combat. This list excluded tasks that generally involved more mental than physical requirements, such as land navigation, first aid, and communicating and reporting. Finally, the research team deconstructed the identified tasks into functional and physical components, with an emphasis on eliminating redundant physical requirements.

Because these five combat-representative tasks cannot be effectively replicated and measured (because of safety reasons, considerable resource requirements, an inability to consistently score, and other factors), the research team assembled subject-matter experts to aid in developing realistic and scorable simulations. The resulting simulations to assess each task were as follows and correspond to the common core tasks on the right side of Figure C.1:

- move to contact: loaded road march
- build hasty fighting position: bucket fill, sandbag stack
- move O-U-A-T (over-under-around-through) obstacles: sprint, crawl, beam walk and carry, obstacle/wall series
- employ progressive levels of force: tire flip, sandbag throw, sled pull, barrel rotation
- extract and move casualty: extricate and drag casualty.

Using the Army's past fitness testing and its broader review of the literature (Hauschild et al., 2014; Knapik et al., 2004, p. 59; Nindl et al., 2015), the BSPRRS team settled on 23 physical fitness events as a comprehensive list for further validation study using soldier samples. These events are shown in Table C.2. The discussions regarding stages 2 and 3 in Chapter Two provide greater detail about these fitness events as predictors and their down-select to the final ACFT events included in the overall battery.

TABLE C.2

Fitness Events Initially Considered for ACFT Development

Fitness Component	Fitness Event	Score Metric
Cardiovascular endurance	Two-mile run	Time (minutes:seconds)
Explosive power	20 lb power throw	Distance (meters)
	50 yd power drag	Time (minutes:seconds)
	50 yd sled push	Time (minutes:seconds)
	Standing long jump	Distance (meters)
	Vertical jump	Distance (meters)
Muscular endurance (core, upper and lower body)	40 lb kettlebell squat	Repetitions (to volitional fatigue)
	Abdominal rower	Repetitions (to volitional fatigue)
	Bench press endurance	Repetitions (to volitional fatigue)
	Dips	Repetitions (to volitional fatigue)
	Modified sit-ups	Repetitions (to volitional fatigue)
	Push-ups	Repetitions (within 2 minutes)
	Sit-ups	Repetitions (within 2 minutes)
	Weighted trunk rotations	Repetitions (to volitional fatigue)
Muscular strength	80 lb sumo squat	Repetitions (to volitional fatigue)
	Bench press strength	Repetitions (to volitional fatigue)
	Hexagon bar deadlift	Repetitions (to volitional fatigue)
	Leg tuck	Repetitions (to volitional fatigue)
	Pull-ups	Repetitions (to volitional fatigue)
Speed and agility	300 m shuttle run	Time (minutes:seconds)
	400 yd sprint	Time (minutes:seconds)
	Illinois shuttle test	Time (minutes:seconds)
	Loaded 300 m shuttle run	Time (minutes:seconds)

SOURCE: Adapted from East, DeGroot, and Muraca-Grabowski, 2019, pp. 18–19.

NOTE: Support for most events (or some variant using weight, distance, or load) has been generally documented in the broader research literature. However, some events (such as the leg tuck, dips, abdominal rower, and weighted trunk rotations) have a history of being part of the broader Army fitness training program and may not have a strong empirical basis for inclusion on this list. The categorization of events into respective fitness components was taken from East, DeGroot, and Muraca-Grabowski, 2019. The definitions of the components, consideration of body regions, and/or inclusion of gender differences could alter the categorization of some events.

APPENDIX D

Differential Prediction Analysis for ACFT Total Score and Individual Event Scores

CIMT provided us with additional information and analysis that the Army conducted to examine the differential prediction of the WTBD-ST by gender. Figure D.1 shows a scatterplot of total ACFT scores relative to the time it took soldiers to complete the WTBD-ST. Figures D.2–D.7 show scatterplots that compare WTBD-ST performance with performance on the six ACFT fitness events.

FIGURE D.1
Differential Prediction Analysis for ACFT Total Score

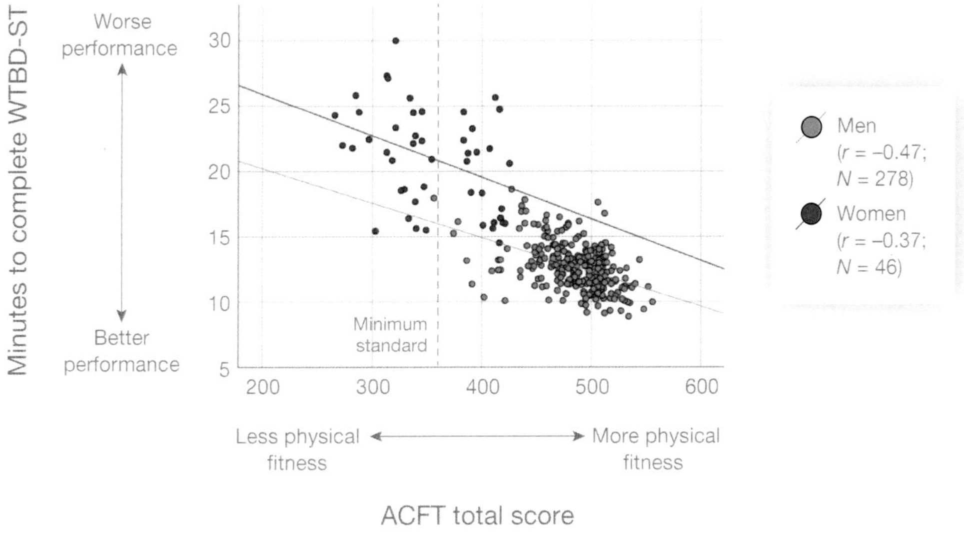

FIGURE D.2
Differential Prediction Analysis for Maximum Deadlift

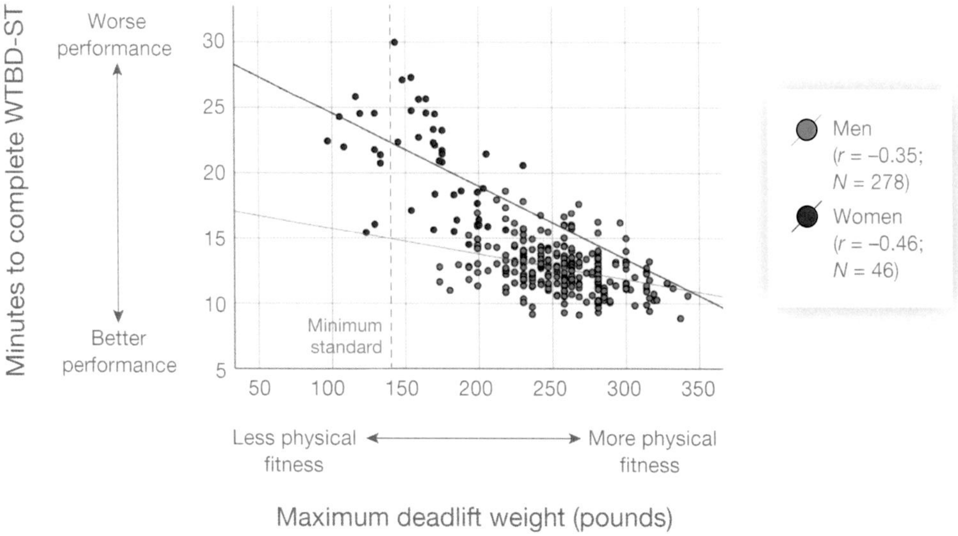

FIGURE D.3
Differential Prediction Analysis for Standing Power Throw

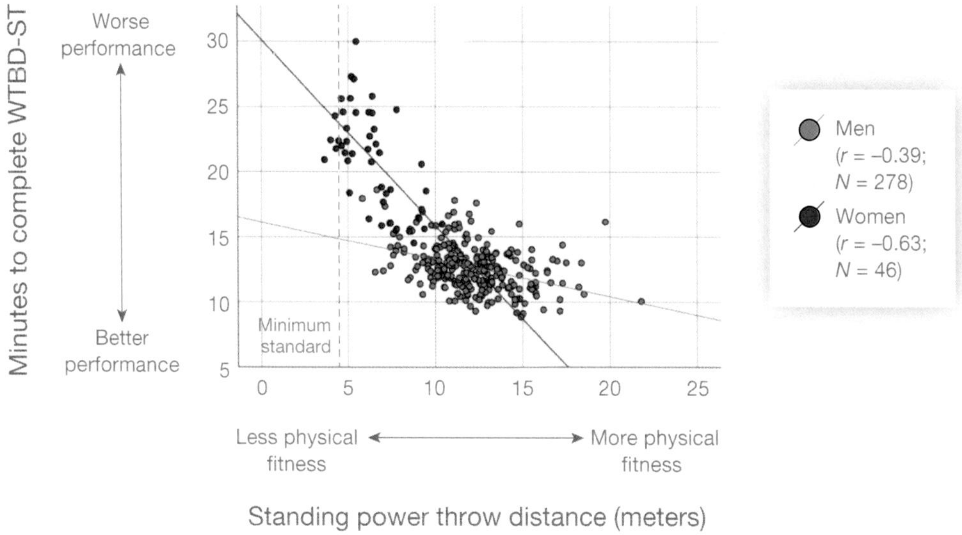

FIGURE D.4
Differential Prediction Analysis for Hand Release Push-Up

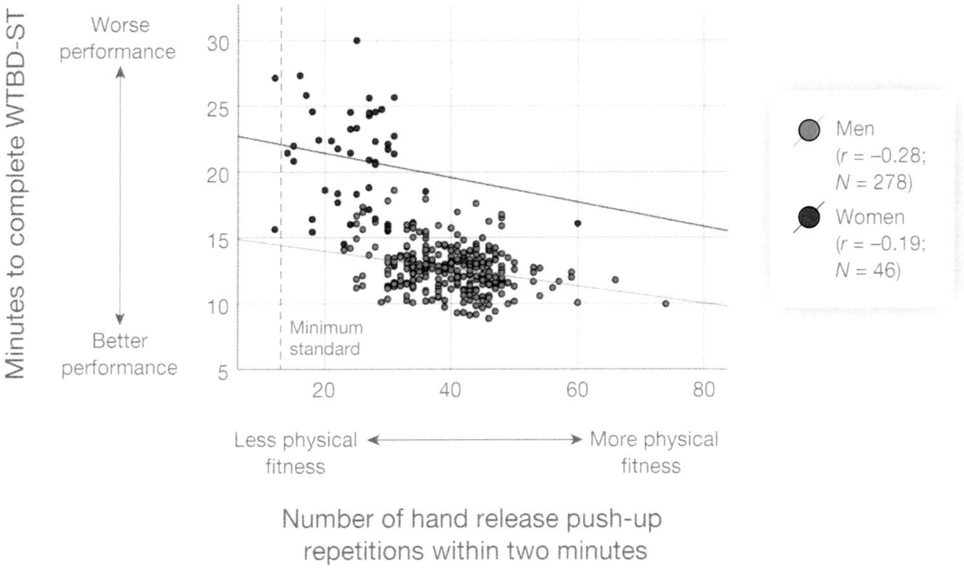

FIGURE D.5
Differential Prediction Analysis for Sprint-Drag-Carry

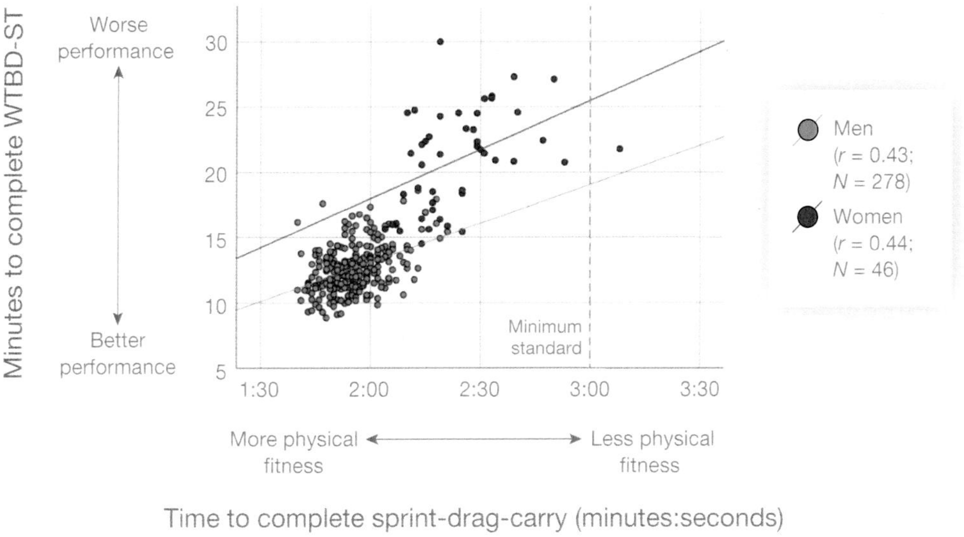

FIGURE D.6
Differential Prediction Analysis for Leg Tuck

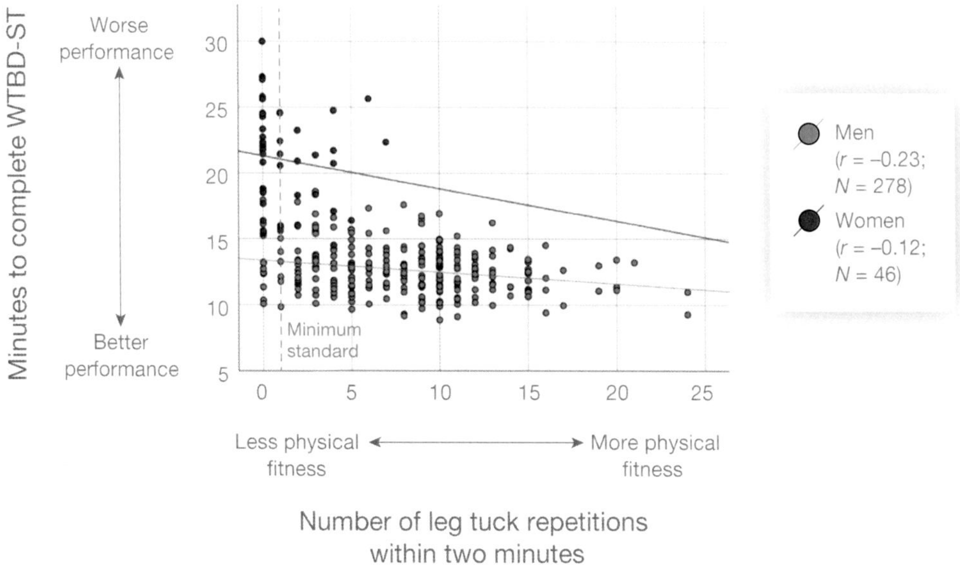

FIGURE D.7
Differential Prediction Analysis for Two-Mile Run

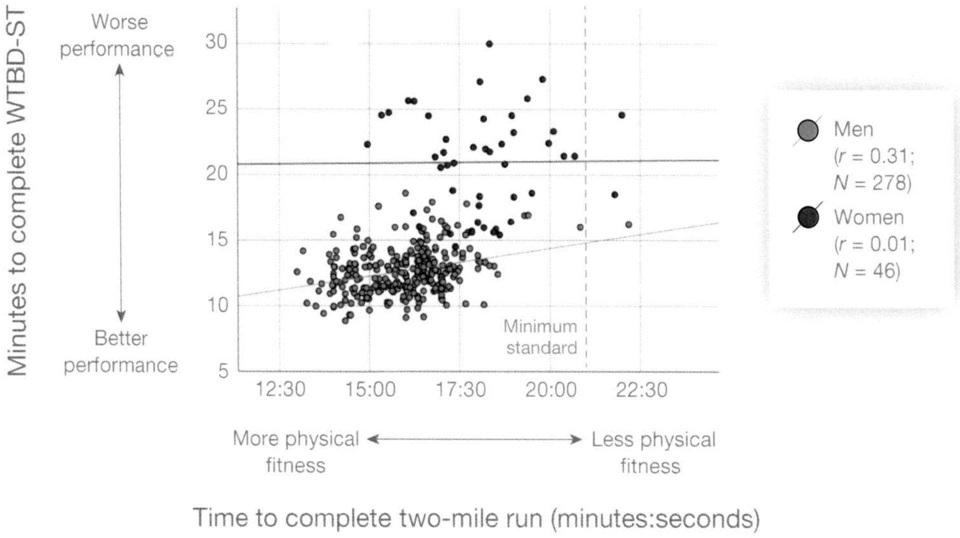

APPENDIX E

Additional Findings on Pass Rates and Event Outcomes

Tables E.1 and E.2 show officer pass rates for the top ten and bottom ten AOCs. These tables correspond to Tables 3.2 and 3.3 for enlisted personnel.

In Table E.3, we show the AOCs that have the lowest pass rates overall. Because some AOCs do not have enough men or women in them to provide stable estimates by gender, we have omitted the by-gender estimates for which the number of soldiers in that cell is too small. However, we have included those AOCs in the table because the overall pass rate was among the ten lowest when considering the data from both genders. Those with omitted information are listed at the top. Those with complete information are listed in order from lowest to highest pass rate.

Table 3.5 showed that enlisted personnel under the age of 45, regardless of component or gender, had similar pass rates on the ACFT but that soldiers over the age of 45 experienced a significant drop-off in pass rates. Table E.4 shows the same information (and patterns) for officers.

Chapter Three showed pass rates by age group. Figure E.1 shows the results of a multivariate regression analysis that examined the relationship between age and event scores. The results of this analysis are similar to the observations made in Chapter Three, in that most event scores dip for the oldest age group. For men ages 44 and older, scores drop by about 5 points on the maximum deadlift and the two-mile run. There also is a sizable drop in scores for both men and women on the sprint-drag-carry that ranges from about 9 points for women to about 12 points for men. The decline in scores starts for both genders around ages 35 to 44.

Figure 4.2 showed which events Regular Army enlisted soldiers were failing. Figure E.2 shows the same results for USAR and ARNG personnel.

TABLE E.1

ACFT Pass Rates for Top Ten and Bottom Ten Areas of Concentration Among Regular Army Officer Women

AOC	Proportion Passing	Number of Women in MOS with ACFT Scores[a]
Top Ten		
19A - ARMOR	93%	84
38A - CIVIL AFFAIRS (AA AND USAR)	91%	64
89E - EXPLOSIVE ORDNANCE DISPOSAL	90%	40
11A - INFANTRY	88%	64
12A - ENGINEER	86%	339
65B - PHYSICAL THERAPY	86%	90
15B - AVIATION COMBINED ARMS OPERATIONS	85%	73
15A - AVIATION, GENERAL	85%	142
31A - MILITARY POLICE	84%	274
35D - ALL SOURCE INTELLIGENCE	82%	433
Bottom Ten		
67B - ALLIED SCIENCES	45%	53
66P - FAMILY NURSE PRACTITIONER	47%	86
60W - PSYCHIATRIST	47%	36
61F - INTERNIST	47%	55
66E - PERIOPERATIVE NURSE	51%	65
60P - PEDIATRICIAN	53%	53
66G - OBSTETRICS AND GYNECOCOLOGY	55%	86
66F - NURSE ANESTHETIST	55%	62
60J - OBSTETRICIAN AND GYNECOLOGIST	56%	78
67E - PHARMACY	57%	35

NOTE: Pass rates are only included in the table if there are at least 30 women tested in the AOC. Pass rates reported here include the most recent test taken from 2019 through September 10, 2021.

[a] Those who completed an alternate event, were listed as being on profile, had missing data on one or more events, or had scores outside the expected range of scores on an event were excluded.

TABLE E.2

ACFT Pass Rates for Top Ten and Bottom Ten Areas of Concentration Among Regular Army Officer Men

AOC	Proportion Passing	Number of Men in MOS with ACFT Scores[a]
Top Ten		
60F - PULMONARY DISEASE/CRITICAL CARE OFFICER	100%	39
51A - SYSTEMS DEVELOPMENT	100%	53
18A - SPECIAL FORCES	100%	303
11A - INFANTRY	99%	4,648
15A - AVIATION, GENERAL	99%	949
89E - EXPLOSIVE ORDNANCE DISPOSAL	99%	339
60J - OBSTETRICIAN AND GYNECOLOGIST	98%	54
15B - AVIATION COMBINED ARMS OPERATIONS	98%	705
19A - ARMOR	98%	1,891
35D - ALL SOURCE INTELLIGENCE	98%	1,467
Bottom Ten		
61N - FLIGHT SURGEON	71%	31
67B - ALLIED SCIENCES	84%	150
61F - INTERNIST	85%	204
61P - PHYSIATRIST	86%	35
60V - NEUROLOGIST	86%	35
60W - PSYCHIATRIST	86%	113
61J - GENERAL SURGEON	88%	123
60N - ANESTHESIOLOGIST	88%	115
56A - COMMAND AND UNIT CHAPLAIN	88%	662
62B - FIELD SURGEON	89%	72

NOTE: Pass rates are only included in the table if there are at least 30 men tested in the AOC. Pass rates reported here include the most recent test taken from 2019 through September 10, 2021.

[a] Those who completed an alternate event, were listed as being on profile, had missing data on one or more events, or had scores outside the expected range of scores on an event were excluded.

TABLE E.3

ACFT Pass Rates for Bottom Ten Areas of Concentration Overall (Regular Army)

	Proportion Passing			Number in MOS with ACFT Scores[a]	
AOC	Overall	Men	Women	Men	Women
66G - OBSTETRICS AND GYNECOLOGY	—	—	55%	≤5	86
61N - FLIGHT SURGEON	—	71%	—	31	≤5
60L - DERMATOLOGIST	—	89%	—	36	19
66P - FAMILY NURSE PRACTITIONER	62%	93%	47%	42	86
66E - PERIOPERATIVE NURSE	72%	92%	51%	73	65
60J - OBSTETRICIAN AND GYNECOLOGIST	73%	98%	56%	54	78
67B - ALLIED SCIENCES	74%	84%	45%	150	53
60P - PEDIATRICIAN	74%	90%	53%	71	53
65C - DIETITIAN	75%	95%	66%	39	95
61U - PATHOLOGIST	75%	90%	43%	49	23

NOTE: Pass rates are only included in the table if there are at least 30 men or 30 women tested in the AOC. Pass rates reported here include the most recent test taken from 2019 through September 10, 2021.

[a] Those who completed an alternate event, were listed as being on profile, had missing data on one or more events, or had scores outside the expected range of scores on an event were excluded.

TABLE E.4

ACFT Pass Rates for Officers, by Age Group

Gender	Component	18–24	25–34	35–44	45-Plus
Men	Regular Army	99%	98%	95%	88%
	USAR	92%	91%	89%	77%
	ARNG	97%	94%	91%	82%
Women	Regular Army	83%	75%	64%	37%
	USAR	—	56%	49%	35%
	ARNG	65%	59%	54%	45%
Overall	Regular Army	95%	93%	90%	82%
	USAR	81%	83%	81%	71%
	ARNG	90%	89%	87%	79%

NOTE: Pass rates reported here include the most recent test taken from 2019 through September 10, 2021. Those who completed an alternate event, were listed as being on profile, had missing data on one or more events, or had scores outside of the expected range of scores on an event were excluded. Data are omitted for cells with fewer than 30 people tested.

FIGURE E.1
Effect of Age on ACFT Event Scores

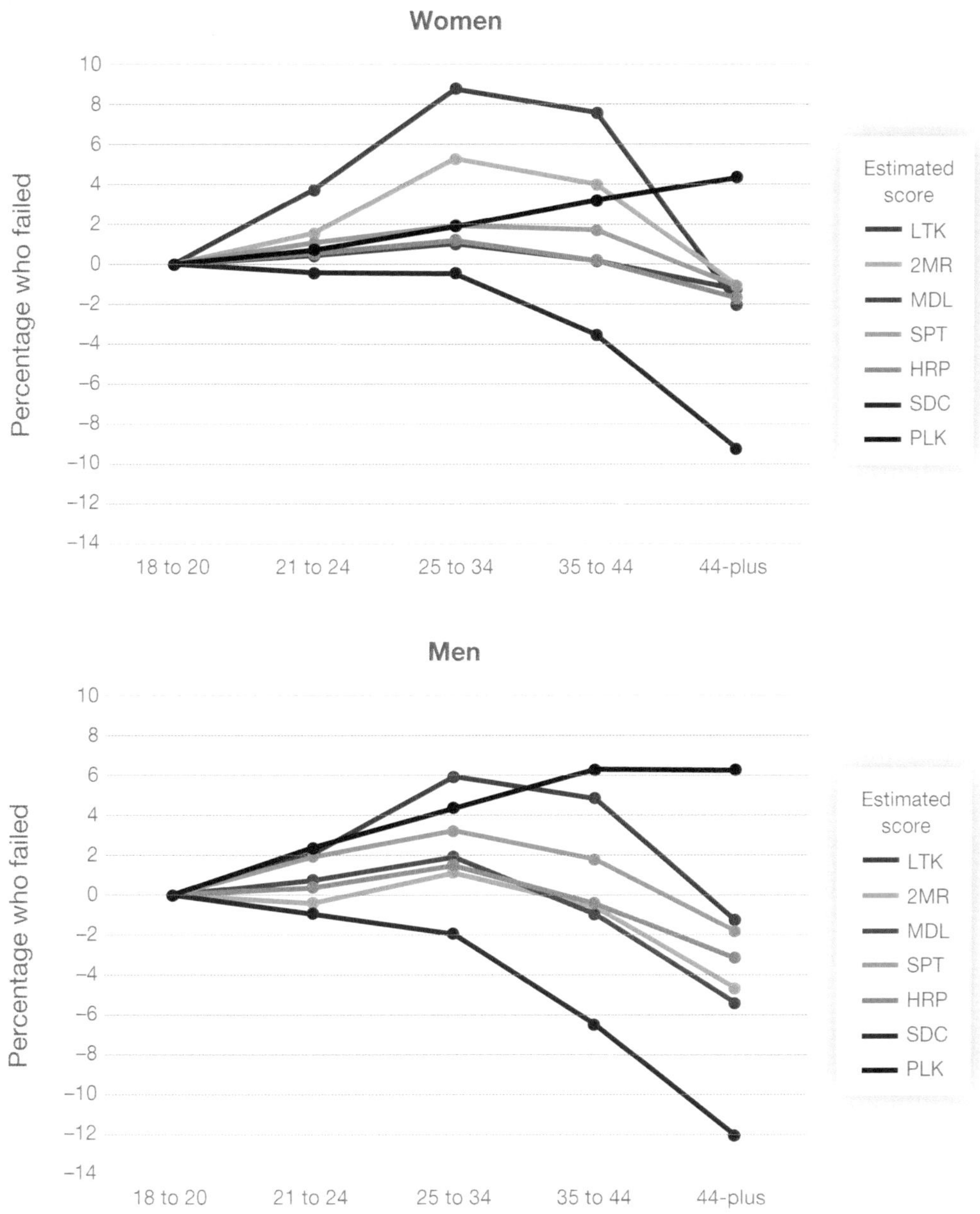

NOTE: LTK = leg tuck; 2MR = two-mile run; MDL = maximum deadlift; SPT = standing power throw; HRP = hand release push-up; SDC = sprint-drag-carry; PLK = plank. This figure shows a regression analysis in which the event score was regressed on the following independent variables: MOS, region of the country, elevation, weather, height, weight, and race and ethnicity. Regressions were run separately for each gender. The figures show the predicted ACFT scores for each age group using the beta weights from the gender-specific regression equation. Age groups were entered as dummy variables to allow for a nonlinear relationship.

FIGURE E.2

Failure Rates by Event for Those Who Failed the ACFT Overall (U.S. Army Reserve and Army National Guard Enlisted)

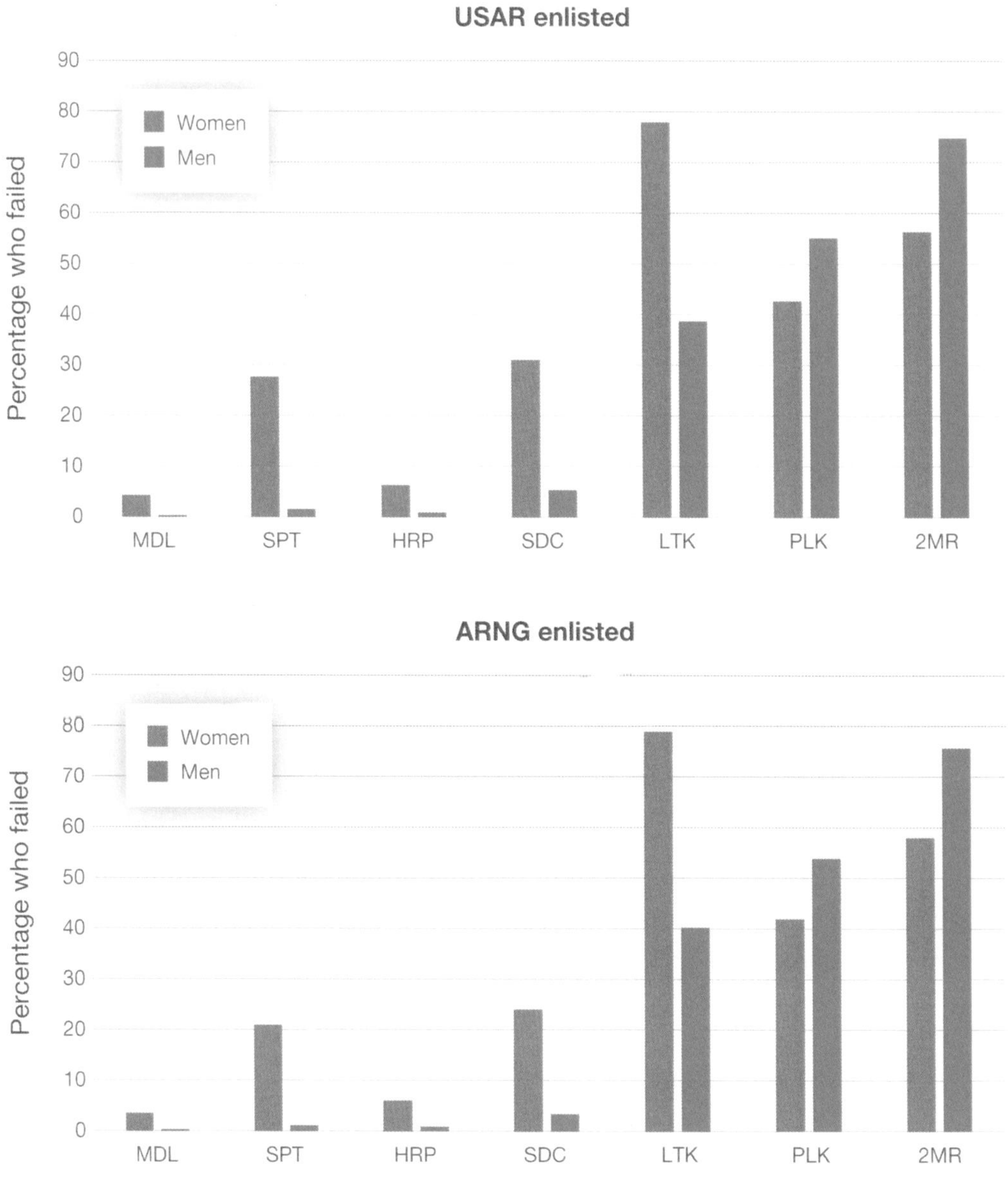

NOTE: MDL = maximum deadlift; SPT = standing power throw; HRP = hand release push-up; SDC = sprint-drag-carry; LTK = leg tuck; PLK = plank; 2MR = two-mile run. Sample sizes are the same for all events except the leg tuck and plank. The number of women overall is 4,223 (USAR) and 7,282 (ARNG), and the number of men is 5,058 (USAR) and 11,151 (ARNG). Data for the leg tuck and plank include soldiers who have a score of zero on both events.

Pass Rates Under Different Policy Options

Fitness Tier Cut Point Analysis for Men

Figure 3.1 showed the proportion of women who would be assigned to each fitness tier—platinum, gold, silver, bronze, and green—under two approaches for establishing cut points: (1) using only passing scores and (2) using all ACFT scores, regardless of pass/fail status. Figure F.1 shows the results of the same analysis for men. Unlike the results for women, there is very little difference for men between the two approaches.

Lowering Event Minimums

As discussed in Chapter Three, one policy option for mitigating differential pass rates would be to lower the minimum score required to pass each event (except the leg tuck, which has a passing requirement of one repetition). Figure F.2 shows pass rates for minimum scores of 40 and 50 for men and women, as well as pass rates if the leg tuck was no longer required. Reducing minimums would improve pass rates for women to some extent, though it would not eliminate all failures. There would not be a significant impact for men because of their high pass rates under the current policy.

FIGURE F.1
Proportion of Regular Army Enlisted Men at Each ACFT Fitness Tier, with Cut Points Calculated Two Ways

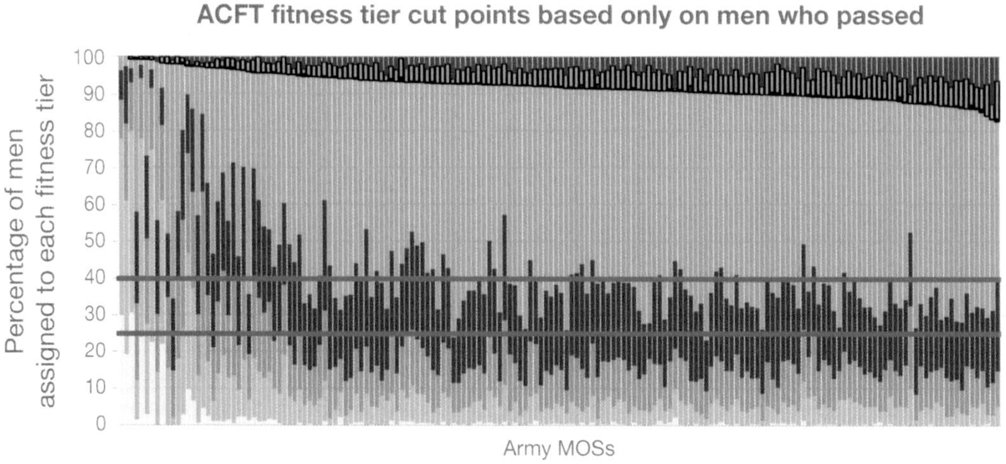

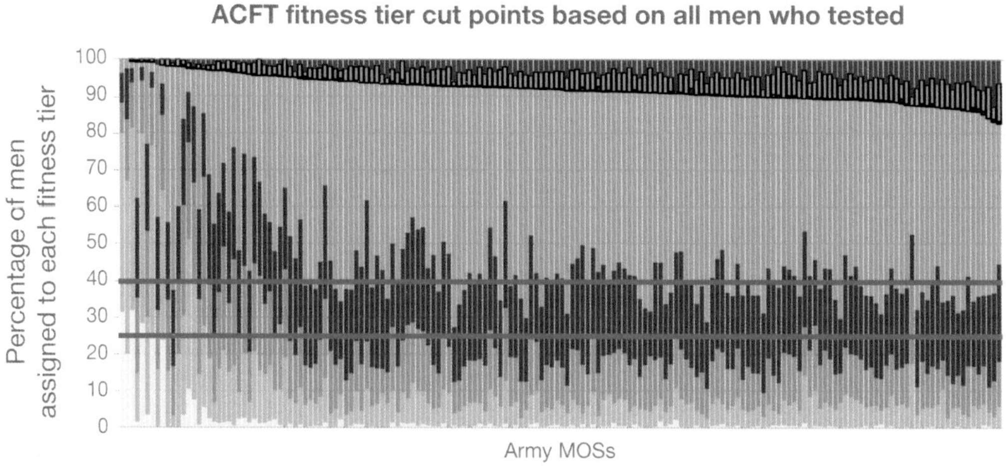

ACFT Tier	Percentile (by gender)	Regular Army Men Overall	Regular Army Men Passing
Platinum	Top 1%	585	586
Gold	Top 10%	543	546
Silver	Top 25%	509	512
Bronze	Top 50%	470	475
Green	360 and up to 50%	360	360
Not Eligible	Below 360		

NOTE: Both figures contain scores from all Regular Army men test-takers. However, tiers (the bottom portion of both figures) are constructed differently according to the two approaches considered. In the top figure, percentiles are calculated using only passing scores. In the bottom figure, all scores are included when calculating percentiles. Bars outlined in black indicate scores of 360 and above in cases in which soldiers failed the ACFT.

FIGURE F.2

Increase in Proportion Passing Under a 50 or 40 Minimum Policy or No Leg Tuck, by Gender (Enlisted Regular Army)

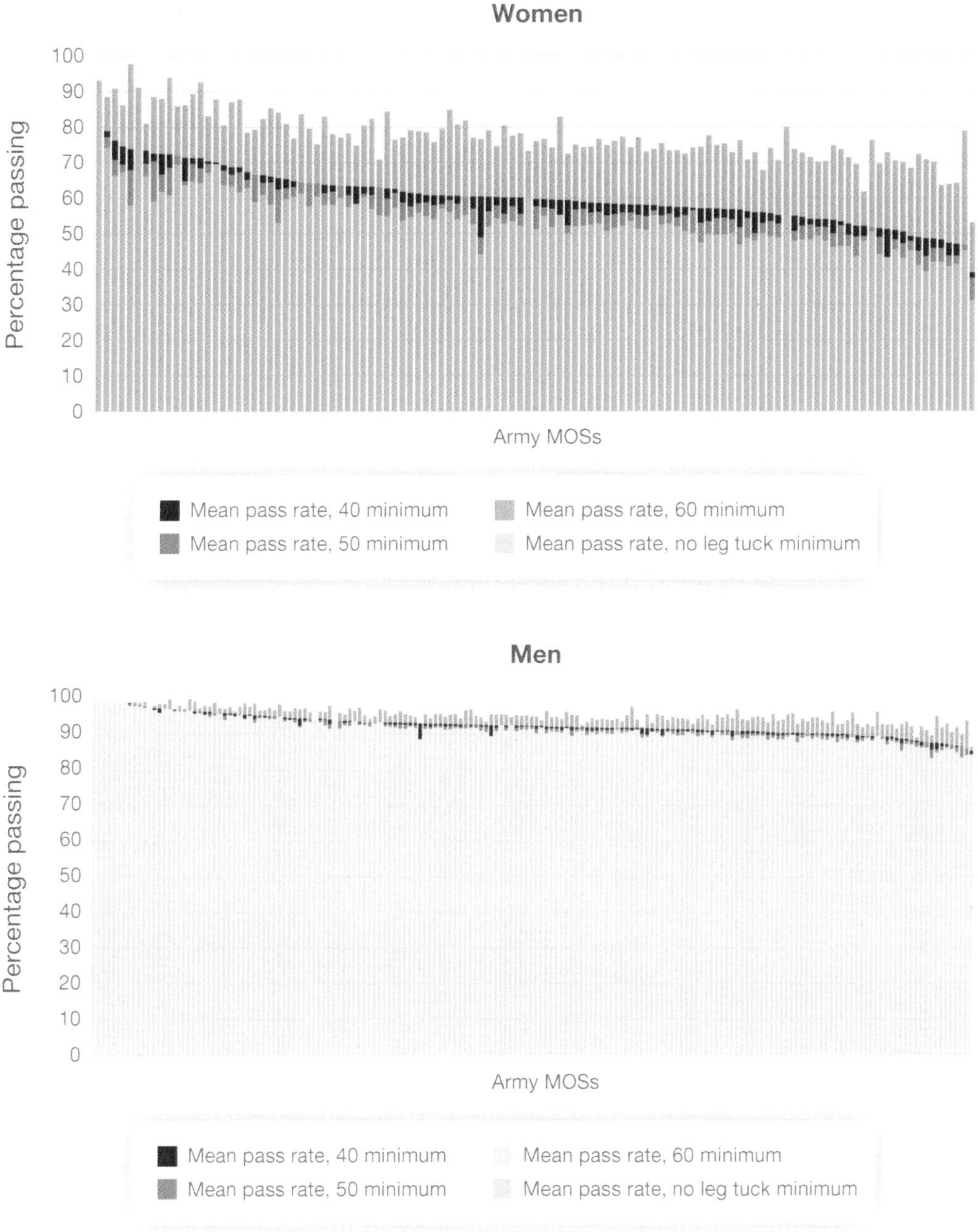

NOTE: Estimates use soldiers' most recent ACFT taken from 2019 through September 10, 2021 (*N* = 244,055 men and 35,388 women). All values were calculated using ACFT 3.0 rules for scoring and pass/fail minimums. Estimates for MOSs with fewer than 30 men or women with an ACFT score are omitted from the figure. Individuals who completed an alternate event, were listed as being on profile, or who did not have a valid score for all six events were also omitted.

Abbreviations

ACFT	Army Combat Fitness Test
AOC	area of concentration
APFT	Army Physical Fitness Test
ARNG	Army National Guard
BCT	Basic Combat Training
BSPRRS	Baseline Soldier Physical Readiness Requirements Study
CIMT	U.S. Army Center for Initial Military Training
COVID-19	coronavirus disease 2019
CST	common soldier task
GAO	U.S. Government Accountability Office
H2F	Holistic Health and Fitness
MFT	master fitness trainer
MOS	military occupational specialty
USAR	U.S. Army Reserve
USMA	U.S. Military Academy
WTBD	warrior task and battle drill
WTBD-ST	warrior task and battle drill simulation test

References

Abdel-Malek, Karim, Laura Frey-Law, Rajan Bhatt, Marco Tena Salais, Chris Murphy, Kaylee Lichtenstein, and Russell Schneider, *Army Combat Fitness Test (ACFT) External Validation: Strength Evaluation of the ACFT—Preliminary Results*, Iowa City, Iowa: Iowa Technology Institute, University of Iowa, 2021, Not available to the general public.

American Educational Research Association, American Psychological Association, and National Council on Measurement in Education, *Standards for Educational and Psychological Testing*, Washington, D.C.: American Educational Research Association, 2014. As of July 1, 2021: https://www.testingstandards.net/uploads/7/6/6/4/76643089/standards_2014edition.pdf

Army Regulation 600-8-24, *Officer Transfers and Discharges*, Washington, D.C.: Headquarters, Department of the Army, February 8, 2020.

Bishop, Phillip, Kirk Cureton, and Mitchell Collins, "Sex Difference in Muscular Strength in Equally-Trained Men and Women," *Ergonomics*, Vol. 30, No. 4, 1987, pp. 675–687.

Brading, Thomas, "ACFT 2.0: Changes Sparked by COVID-19," U.S. Army, June 15, 2020. As of December 14, 2020:
https://www.army.mil/article/236479/acft_2_0_changes_sparked_by_covid_19

Callinan, Militza, and Ivan T. Robertson, "Work Sample Testing," *International Journal of Selection and Assessment*, Vol. 8, No. 4, 2000, pp. 248–260.

Center for Initial Military Training, U.S. Army Training Command, "ACFT 3.0," webpage, March 23, 2021. As of December 14, 2021:
https://www.army.mil/standto/archive/2021/03/23/

Courtright, Stephen H., Brian W. McCormick, Bennett E. Postlethwaite, Cody J. Reeves, and Michael K. Mount, "A Meta-Analysis of Sex Differences in Physical Ability: Revised Estimates and Strategies for Reducing Differences in Selection Contexts," *Journal of Applied Psychology*, Vol. 98, No. 4, 2013, pp. 623–641.

Dada, Esther O., Morgan K. Anderson, Tyson Grier, Joseph A. Alemany, and Bruce H. Jones, "Sex and Age Differences in Physical Performance: A Comparison of Army Basic Training and Operational Populations," *Journal of Science and Medicine in Sport*, Vol. 20, Supp. 4, 2017, pp. S68–S73.

de la Motte, Sarah J., Timothy C. Gribbin, Peter Lisman, Kaitlin Murphy, and Patricia A. Deuster, "Systematic Review of the Association Between Physical Fitness and Musculoskeletal Injury Risk: Part 2—Muscular Endurance and Muscular Strength," *Journal of Strength and Conditioning Research*, Vol. 31, No. 11, 2017, pp. 3218–3234.

de la Motte, Sarah J., Peter Lisman, Timothy C. Gribbin, Kaitlin Murphy, and Patricia A. Deuster, "Systematic Review of the Association Between Physical Fitness and Musculoskeletal Injury Risk: Part 3—Flexibility, Power, Speed, Balance, and Agility," *Journal of Strength and Conditioning Research*, Vol. 33, No. 6, 2019, pp. 1723–1735.

East, Whitfield B., David DeGroot, and Stephanie Muraca-Grabowski, *Baseline Soldier Physical Readiness Requirements Study*, Fort Eustis, Va.: U.S. Army Center for Initial Entry Training, Technical Report T19.041-13.1, November 2019. As of June 22, 2021:
https://www.iadlest.org/Portals/0/AD1097586%20Baseline%20Soldier%20Physical%20Readiness%20Requirements%20Study.pdf?ver=lI8Q0CmCEx_rmeabYlwG4Q%3D%3D

Field Manual 7-22, *Holistic Health and Fitness*, Washington, D.C.: Headquarters, Department of the Army, October 2020.

GAO—*See* U.S. Government Accountability Office.

Grier, Tyson, Timothy Benedict, Bruce Jones, and Michelle Chervak, *ACFT H2F Evaluation: Part I—Injury, Training and Human Performance*, Aberdeen Proving Ground, Md.: U.S. Army Public Health Center, March 9, 2021, Not available to the general public.

Hardison, Chaitra M., Susan D. Hosek, and Chloe E. Bird, *Establishing Gender-Neutral Physical Standards for Ground Combat Occupations: Volume 1. A Review of Best-Practice Methods*, Santa Monica, Calif.: RAND Corporation, RR-1340/1-OSD, 2018. As of June 12, 2021: https://www.rand.org/pubs/research_reports/RR1340z1.html

Hauschild, Veronique, David DeGroot, Shane Hall, Karen Deaver, Keith Hauret, Tyson Grier, and Bruce Jones, *Correlations Between Physical Fitness Tests and Performance of Military Tasks: A Systematic Review and Meta-Analyses*, Aberdeen Proving Ground, Md.: U.S. Army Public Health Command, PHR No. 12-02-0614, June 2014. As of January 25, 2022: https://apps.dtic.mil/sti/citations/ADA607688

Knapik, Joseph J., Bruce H. Jones, Marilyn A. Sharp, Salima Darakjy, Sarah Jones, Keith G. Hauret, and Gene Piskator, *The Case for Pre-Enlistment Physical Fitness Testing: Research and Recommendations*, Aberdeen Proving Ground, Md.: U.S. Army Center for Health Promotion and Preventive Medicine, No. 12-HF-01Q9D-04, August 2004. As of January 25, 2022: https://apps.dtic.mil/sti/citations/ADA426848

Kraemer, William J., Scott A. Mazzetti, Bradley C. Nindl, Lincoln A. Gotshalk, Jeff S. Volek, Jill A. Bush, Jim O. Marx, Kei Dohi, Ana L. Gómez, Mary Miles, Steven J. Fleck, Robert U. Newton, and Keijo Häkkinen, "Effect of Resistance Training on Women's Strength/Power and Occupational Performances," *Medicine and Science in Sports and Exercise*, Vol. 33, No. 6, 2001, pp. 1011–1025.

Lisman, Peter J., Sarah J. de la Motte, Timothy C. Gribbin, Dianna P. Jaffin, Kaitlin Murphy, and Patricia A. Deuster, "A Systematic Review of the Association Between Physical Fitness and Musculoskeletal Injury Risk: Part 1—Cardiorespiratory Endurance," *Journal of Strength and Conditioning Research*, Vol. 31, No. 6, 2017, pp. 1744–1757.

McGill, Stuart, David Frost, Thomas Lam, Tim Finlay, Kevin Darby, and Jordan Cannon, "Can Fitness and Movement Quality Prevent Back Injury in Elite Task Force Police Officers? A 5-Year Longitudinal Study," *Ergonomics*, Vol. 58, No. 10, 2015, pp. 1682–1689.

Meredith, Lisa S., Carra S. Sims, Benjamin Saul Batorsky, Adeyemi Okunogbe, Brittany L. Bannon, and Craig A. Myatt, *Identifying Promising Approaches to U.S. Army Institutional Change: A Review of the Literature on Organizational Culture and Climate*, Santa Monica, Calif.: RAND Corporation, RR-1588-A, 2017. As of October 31, 2021: https://www.rand.org/pubs/research_reports/RR1588.html

Motowidlo, Stephan J., Marvin D. Dunnette, and Gary W. Carter, "An Alternative Selection Procedure: The Low-Fidelity Simulation," *Journal of Applied Psychology*, Vol. 75, No. 6, 1990, pp. 640–647.

Muraca-Grabowski, Stephanie, *2019 Physical Training and Readiness Survey Report*, Fort Eustis, Va.: Center for Initial Military Training, 2020, Not available to the general public.

Nindl, Bradley C., "Physical Training Strategies for Military Women's Performance Optimization in Combat-Centric Occupations," *Journal of Strength and Conditioning Research*, Vol. 29, Supp. 11, 2015, pp. S101–S106.

Nindl, Bradley C., Brent A. Alvar, Jason R. Dudley, Mike W. Favre, Gerard J. Martin, Marilyn A. Sharp, Brad J. Warr, Mark D. Stephenson, and William J. Kraemer, "Executive Summary from the National Strength and Conditioning Association's Second Blue Ribbon Panel on Military Physical Readiness: Military Physical Performance Testing," *Journal of Strength and Conditioning Research*, Vol. 29, Supp. 11, November 2015, pp. S216–S220.

Pierce, Joseph R., Keith G. Hauret, Joseph A. Alemany, Tyson L. Grier, Marilyn A. Sharp, Jan E. Redmond, Stephen A. Foulis, Bruce S. Cohen, Maria C. Canino, and Bruce H. Jones, *Physical Performance on the Occupational Physical Assessment Test (OPAT), Army Physical Fitness Test (APFT), and Relationship to Body Mass Index During Initial Entry Training—OPAT Phase I*, Aberdeen Proving Ground, Md.: U.S. Army Public Health Center, PHR No. S.0047229-18a, August 2018. As of July 1, 2021:
https://apps.dtic.mil/sti/pdfs/AD1058052.pdf

Public Law 116-283, William M. (Mac) Thornberry National Defense Authorization Act for Fiscal Year 2021, Section 598, Limitation on Implementation of Army Combat Fitness Test, January 1, 2021.

Society for Industrial and Organizational Psychology, *Principles for the Validation and Use of Personnel Selection Procedures*, 5th ed., Bowling Green, Ohio, August 2018. As of July 1, 2021:
https://www.apa.org/ed/accreditation/about/policies/personnel-selection-procedures.pdf

Soldier Training Publication 21-1-SMCT, *Soldier's Manual of Common Tasks: Warrior Skills Level 1*, Washington, D.C.: Headquarters, Department of the Army, November 2019.

U.S. Army, "Army Combat Fitness Test," webpage, undated. As of January 25, 2022:
https://www.army.mil/acft/

———, *U.S. Army Holistic Health and Fitness Operating Concept: The U.S. Army's System for Enhancing Soldier Readiness and Lethality in the 21st Century*, Washington, D.C., October 1, 2020. As of January 5, 2022:
https://www.army.mil/e2/downloads/rv7/acft/h2f_operating_concept.pdf

U.S. Army Fort Benning and the Maneuver Center of Excellence, "ARNG Warrior Training Center: Master Fitness Trainer Course," webpage, updated July 21, 2021. As of December 13, 2021:
https://www.benning.army.mil/Tenant/WTC/MFTC.html

U.S. Army Public Health Center, *Annual Injury Surveillance Report 2019 Summary*, Aberdeen Proving Ground, Md., Technical Information Paper No. 12-114-0121, December 1, 2020. As of January 25, 2022:
https://apps.dtic.mil/sti/citations/AD1136242

U.S. Government Accountability Office, *Gender Issues: Improved Guidance and Oversight Are Needed to Ensure Validity and Equity of Fitness Standards*, Washington, D.C., GAO/NSIAD-99-9, November 1998. As of June 12, 2021:
https://www.gao.gov/assets/nsiad-99-9.pdf

———, *Military Personnel: Additional Actions Needed to Address Gaps in Military Physician Specialties*, Washington, D.C., GAO-18-77, February 2018. As of January 11, 2022:
https://www.gao.gov/assets/gao-18-77.pdf

———, *Defense Health Care: Actions Needed to Define and Sustain Wartime Medical Skills for Enlisted Personnel*, Washington, D.C., GAO-21-337, June 2021. As of January 11, 2022:
https://www.gao.gov/assets/gao-21-337.pdf

Varley-Campbell, Jo, Chris Cooper, Daryl Wilkerson, Sophie Wardle, Julie Greeves, and Theo Lorenc, "Sex-Specific Changes in Physical Performance Following Military Training: A Systematic Review," *Sports Medicine*, Vol. 48, No. 11, 2018, pp. 2623–2640.

Young, John W., *Differential Validity, Differential Prediction, and College Admission Testing: A Comprehensive Review and Analysis*, New York: College Entrance Examination Board, No. 2001-6, 2001.